A Comprehensive Method for Concrete Mix Design

Author
Kambiz Janamian

Co-author
Jose Aguiar

I dedicate this to:
*my beloved wife Parinaz, the best participant in my life,
and my little son*

Published by **Materials Research Forum LLC**
Millersville, PA 17551, USA

Published as part of the book series
Materials Research Foundations
Volume 65 (2020)
ISSN 2471-8890 (Print)
ISSN 2471-8904 (Online)

Print ISBN 978-1-64490-058-1
ePDF ISBN 978-1-64490-059-8

Distributed worldwide by

Materials Research Forum LLC
105 Springdale Lane
Millersville, PA 17551
USA
http://www.mrforum.com

Printed in the United States of America
10 9 8 7 6 5 4 3 2 1

Table of Contents

1. Introduction

The concrete mix design is one of the most important challenges in construction projects. The use of high quality and high-performance concrete is growing every day in the world, so the improvement of knowledge of concrete mix design with high and new technology is getting more important.

There are a lot of different methods for concrete mix design in texts and documents. Most of the time, users cannot decide which method to use to get the best results. On the other hand, when we use any of these methods, final results will likely not be compatible with the real industry. When we design a concrete with any method, after making laboratory trials, we need to produce a concrete in a real batching plant. There have been many different samples where we used an in the lab approved mix design in a batching plant, but unfortunately the final industrial results were not compatible with the desired defined properties of the concrete.

Another issue is the knowledge of concrete technology in construction industry. In fact, there is often not enough experience in the case of concrete technology and mix design for engineers graduating from universities. Students often only pass some courses in the university and of course, this is not enough experience for concrete technology and mix design. So, when they come across a case of a concrete mix design in a real project, it usually is a big challenge for them.

In this book, we will use a special and novel method for the concrete mix design. This novel method is a composition of many methods of concrete mix design. First, we will study the quality control and specification of concrete constituent materials since we should have good knowledge of these materials for the mix design. After that we will study the specifications of concrete, especially the cases for which we need to confirm our mix design method. Then we will study the novel mix design method and explain the method step by step. Finally, we will test the method with lab trials and after the lab trials we will produce different concrete mixes in a real batching plant and test the mixes in the real industry.

We use this method for many different concrete mixes from C25 to C90 normal weight concretes. We use local aggregates and cement from areas nearby. We also put different concrete mixes in truck mixers and transport them to the final places for real projects. There we take samples of the concrete from the concrete mixer truck and measure different specification like slump and compressive strength. Using this procedure, we can evaluate our mix design method.

To achieve a good and acceptable mix design method, we did many tests in the laboratory and also in real ready mixed concrete plants. First we used a simplified method. After each series of tests, we made some changes in the method and modified that by considering our test results.

At the end we arrive at an acceptable and precise mix design method which works with all kinds of normal constituent materials.

2. Concrete constituent materials specification

Before we can work on the subject of mix design, it is important to have specific information about the concrete constituent materials because the concrete specification depends on these materials properties. So, it is important to do some experiments on these materials to achieve defined properties. In some other cases, we need to get some information from the producer of a material. For some of the materials like cement and super-plasticizers, one needs to test them very precisely to acquire the exact information needed.

In this part of the book, we will study all constituent materials and will explain the tests that we have to do to get the results.

To produce a stable concrete, it is very important to control the quality and stability of all these constituent materials during the production process. For example, if we use a kind of concrete for a project and the project is going to last twelve months, we are supposed to control the quality and stability during these twelve months to achieve the same results. But first, it is important to have a good mix design to make a concrete which satisfies our defined specification.

2.1 Portland cement

Portland cement is the most important constituent material in any kind of concrete because the hardening process of concrete is the result of hydration reaction between the cement and water. We can explain the most important part of this reaction through the following equations:

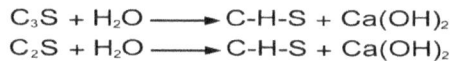

$$C_3S + H_2O \longrightarrow C\text{-}H\text{-}S + Ca(OH)_2$$
$$C_2S + H_2O \longrightarrow C\text{-}H\text{-}S + Ca(OH)_2$$

In these equations C_3S and C_2S are the most important constituents of Portland cement, and the C-H-S is the operative of strength and the durability in concrete.

By considering what was mentioned we can realize that having enough information about the specification of the Portland cement has a big effect on the accuracy of the mix design. So, it is important to do some tests on the Portland cement to gather such specifications. You can see the most important tests for cement from the ASTM standards [1] in table 2.1.

Table 2.1: Important Portland cement tests from ASTM standards	
ASTM C109	Compressive strength of standard cement mortar
ASTM C114	Chemical analysis of cement clinker
ASTM C150	Standard specifications for all kinds of Portland cement
ASTM C151	Expansion of cement in the autoclave
ASTM C186	Hydration heat of cement
ASTM C188	Specific gravity of cement particles
ASTM C191	Setting time of cement paste
ASTM C204	Fineness of cement by Blain method

For the concrete mix design, it is not important to do all the tests yourself because cement producers have all these test results. So, one can simply request such information.

From cement tests, compressive strength of cement mortar and specific gravity of cement particles are necessary for concrete mix design. Doing exact compressive strength tests in a concrete laboratory is quite difficult because the environment temperature and humidity are very important in this test. We also need a special test machine for 5x5 cm cube specimens. Most of the time, we cannot find this kind of test machine in a normal concrete lab. So, we should ask the results from cement producer. As mentioned before, quality control system of cement producers do this test every day. We need to get the results of about 20 days for minimum value of compressive strength of any cement.

Although doing this test is difficult, we did it for two kinds of cements we would like to use for our mix design control. In figure 2.1 you can see the specimens we used.

Figure 2.1: Cement specimens for ASTM C109 test

---

[1] The ASTM standard is the approved standard in the USA and also in Iran, the place which we tested our mix design method.

2.1.1 Cements for our mix design check

To test our mix design check, we decided to use two types of Portland cement from two different producers. In this part, we will discuss the test results and specification for these two Portland cements.

ASTM Type II Portland cement from company (A)

You can see the specification from quality control of producer in table 2.2 and 2.3.

Table 2.2: Chemical analysis of cement type II from company A derived from producer QC lab

Material	Amount	Standard Value (ISIRI 389)
SiO_2	22%	Min 20%
Al_2O_3	5%	Max 6%
Fe_2O_3	3.82%	Max 6%
CaO	64%	
MgO	1.9%	Max 5%
SO_3	1.5%	Max 3%
K_2O	0.49%	
Na_2O	0.25%	
Cl^-	0.019%	
Insoluble residue	0.46%	Max 0.75%
L.O.I	1.0%	Max 3%
L.S.F	91.0%	
C_3A	6.5%	$5-8$%
Free CaO	1.2%	

Table 2.3: Physical properties of cement type II from company A derived from producer QC lab

Specification	Result	Standard Value (ISIRI 389)
Initial setting time	95 min	Minimum 45 min
Final setting time	150 min	Maximum 360 min
Fineness (Blain)	3050 cm^2/gr	Minimum 2800 cm^2/gr
Autoclave expansion	0.14%	Maximum 0.8%
3 days compressive strength	172 kg/cm^2	Minimum 100kg/cm^2
7 days compressive strength	302 kg/cm^2	Minimum 175kg/cm^2
28 days compressive strength	471 kg/cm^2	Minimum 315kg/cm^2
Particles specific gravity	3.16 kg/L	

As mentioned before, the most important information of cement for mix design is the 28 days compressive strength and particles specific gravity. From table 2.3 we can find that the specific gravity of particles for this cement is 3.16kg/L. So, we will use this result for our mix design check in the future.

For 28 days, the compressive strength result in table 2.3 is the mean value of results derived from one month testing of cement with minimum two samples a day. So, we can use $471 kg/cm^2$ for our mix design. But we ourselves tested two samples of this cement as well to compare it with $471 kg/cm^2$. The results are shown in tables 2.4 and 2.5.

Table2.4: Test results for company (A) type II cement (Sample Code:10222)			
Sample age (Days)	Area (cm²)	Pressure (kg)	Compressive strength (kg/cm²)
7	25	9120	365
7	25	9100	364
28	25	11560	462
28	25	11590	464

Table2.5: Test result for company (A) type II cement (Sample Code:20222)			
Sample age (Days)	Area (cm²)	Pressure (kg)	Compressive strength (kg/cm²)
7	25	9350	374
7	25	9330	373
28	25	11610	464
28	25	11600	464

It can be seen from table 2.4 and 2.5 that the result for compressive strength derived from producer QC lab is correct. So, we will use data from table 2.6 for company (A) cement type II.

Table (2.6): Results for Company (A) cement type II checking the mix design	
Value	Result
Particles specific gravity	3.16 kg/L
28 days compressive strength for 5x5cm cubes	470 kg/cm²

ASTM Type I-525 Portland cement from company (B)

This is the second pure Portland cement that we would like to use for our mix design check. You can see the specification from the quality control of producer in tables 2.7 and 2.8.

Table 2.7: Chemical analysis of cement type I-525 from company (B) derived from producer QC lab

Material	Amount	Standard Value (ISIRI 389)
SiO2	21.4%	
Al2O3	5.3%	
Fe2O3	3.7%	
CaO	63.5%	
MgO	1.6%	Max 5%
SO3	2.7%	Max 3%
K2O	0.32%	
Na2O	0.24%	
L.O.I	1.6%	Max 3%
L.S.F	89.8%	
Free CaO	1.5%	
C3S	47.4%	
C2S	25.7%	
C3A	7.7%	
C4AF	11.3%	

Table 2.8: Physical properties of cement type I-525 from company (B) derived from producer QC lab

Specification	Result	Standard Value (ISIRI 389)
Initial setting time	110 min	Minimum 45 min
Final setting time	160 min	Maximum 360 min
Fineness (Blain)	3710 cm2/gr	Minimum 2800 cm2/gr
Autoclave expansion	0.02%	Maximum 0.8%
2 days compressive strength	213 kg/cm2	Minimum 100kg/cm2
7 days compressive strength	427 kg/cm2	
28 days compressive strength	557 kg/cm2	Minimum 525kg/cm2
Particles specific gravity	3.15 kg/L	

In this case, we tested the compressive strength of cement as well. The results are as shown in table 2.9 and 2.10.

Table 2.9: Test result for company (B) type I-525 cement (Sample Code:10223)

Sample age (Days)	Area (cm2)	Pressure (kg)	Compressive strength (kg/cm2)
7	25	11900	476
7	25	11880	475
28	25	14700	588
28	25	14400	576

Table 2.10: Test result for company (B) type I-525 cement (Sample Code:20223)

Sample age (Days)	Area (cm2)	Pressure (kg)	Compressive strength (kg/cm2)
7	25	12010	480
7	25	12040	482
28	25	13460	538
28	25	13470	539

You can see from table 2.9 and 2.10 that the result for compressive strength derived from producer QC lab is correct. So, we will use data from table 2.11 for company (B) cement type I-525.

Table 2.11: Results for company (b) cement type I-525 to check the mix design

Value	Result
Particles specific gravity	3.15 kg/L
28 days compressive strength for 5x5cm cubes	550 kg/cm2

2.2   Other binders (Supplementary cementitious materials)

The use of other binders such as silica fume, ground granulated blast furnace slag (GGBS), fly ash and natural pozzolans are increasing day by day. We call these materials powdered additives. All kinds of concrete in the world now are made with the use of such materials. The most important reasons for using such materials are their positive effect on the durability for the concrete structures. It is very important to build a durable concrete structure due to sustainable development. We can describe the effect of these materials through the following equations:

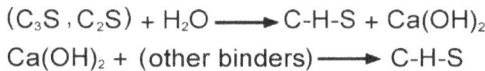

$$(C_3S , C_2S) + H_2O \longrightarrow C\text{-}H\text{-}S + Ca(OH)_2$$
$$Ca(OH)_2 + (other\ binders) \longrightarrow C\text{-}H\text{-}S$$

As you can see, C_3S and C_2S reacts with water and produces C-H-S and $Ca(OH)_2$. In the pozzolanic materials, we have high amount of active SiO_2. If SiO_2 reacts with $Ca(OH)_2$, then we have more C-H-S which is the reason for high strength and low permeability and finally high durability.

By using these materials, we will have more strength and more compressed microstructure in the concrete, but not in the early ages. First, the hydration reaction should happen, then with the production of $Ca(OH)_2$ and active SiO_2 in the pozzolanic materials, the second reaction will happen. So, the action of other binders in the concrete depends on its activity, but it will start at least after seven days and will continue till ninety days or longer.

Now we are going to talk about different kinds of these active powders and use them in concrete.

2.2.1 Silica Fume

Silica fume is a byproduct of ferro-alloy factories which is released from the kiln. It contains more than 90% of active SiO_2. So, it is the most powerful pozzolan. We can call that a super-pozzolan.

For concrete mix design, we need the specific gravity of silica fume that is about 2.2 to 2.3 kg/L. Also, we need to decide for the amount of silica fume in the concrete. You can see suggestions for the amount of silica fume in table 2.12. The suggestions in this table are common uses of all kinds of pozzolanic materials in Asia. So, in other parts of the world, for example European countries, it can be more or less.

For mix design checking, we have access to two producers for silica fume. The quality of both producers are the same. So, we decided to use silica fume from Ferro Silica Company. The specification of this silica fume is in tables 2.13 and 2.14.

Table 2.12: Recommended amount of use for binders in concrete

Material	Recommended from EN 197	Recommended from ACI 211.1	Common use in concrete industry
Silica fume	6% to 10%	5% to 15%	5% to 12%
GGBS	20% to 95%	25% to 70%	20% to 50%
Fly ash	6% to 35%	15% to 35%	10% to 35%
Natural pozzolans	6% to 35%	10% to 20%	10% to 30%

Table 2.13: Chemical properties of silica fume from Ferro Silica Company

SiO2	Fe2O3	CaO	Al2O3	MgO	C	L.O.I	Moisture
Min 85%	Max 2%	Max 1.5%	Max 1%	Max 2%	Max 3%	Max 3.5%	Max 1%

Table 2.14: Physical properties of silica fume from Ferro Silica Company

Structure	Particles shape	Particles dimension	Fineness (Blain)	Specific gravity
Amorphous	Spherical	<40μm	17000cm2/gr	2.25 kg/L

2.2.2 *Ground Granulated Blast Furnace Slag (GGBS)*

GGBS is a byproduct of steel factories. To produce a high quality GGBS, one should use a water jet for cooling slag as soon as it comes out of the kiln. If the cooling process is immediately done, the pozzolanic activity of GGBS is acceptable because in this case the silica in the slag will be transferred to a glass phase and will be activated. Of course if the transfer to glass phase does not happen, the quality of GGBS will not be good enough to use it in concrete for its pozzolanic activity.

For GGBS, we also need the specific gravity for concrete mix design, which is about 2.9 kg/L. As before you can find suggested amounts of use in table 2.12.

We have access to good quality GGBS from steel factories. Some cement producers use this GGBS to produce slag cement with different percentages of slag combined with Portland cement clinker. Unfortunately, some of them do not use separate grinding for the grinding and mixing process of this cement. They feed Portland cement clinker and slag together in to the grinder. In this case, the quality of slag cement can be poor.
We used grinded GGBS from a local Cement Company. The analysis of this GGBS is in the table 2.15.

Table 2.15: GGBS analysis from a local cement company										
SiO2	Cao	Al2O3	MgO	MnO	FeO	TiO2	V2O5	Na2O	K2O	S	H2O
40%	35%	8%	7%	1%	1%	2%	0.2%	0.5%	0.5%	1%	3.3%

The specific gravity of this GGBS is 2.85kg/L. So, in future we will use it for our mix design checking.

2.2.3 *Fly Ash*

Fly ash is a byproduct of power plants which uses collier as its fuel. In ASTM standard, we have two types of fly ash that depend on the type of collier used in the power plant, type C and F. Their specification is somewhat different.

For fly ash like other cementitious materials, we need specific gravity and amount of use for the concrete mix design. The specific gravity for fly ash is about 1.9 to 2.8 kg/L and the amount of use is listed in table 2.12.

Unfortunately, we don't have access to good quality fly ash locally. So, we cannot check our mix design process with fly ash.

2.2.4 *Natural Pozzolans*

Natural pozzolans are natural materials with active SiO_2. So, they have pozzolanic activity. These materials are usually volcanic based or from calcined clay or meta kaoline. Like other binders, these materials can also improve concrete properties.

For concrete mix design, we need natural pozzolans specific gravity which is about 2.4 to 2.7 kg/L and the amount of use listed in table 2.12.

We have access to natural pozzolans. But they locally use it in the production of blended pozzolan cements. So, we couldn't find any separate ground natural pozzolan to use and check our mix design process. You should also know that the specification of any natural pozzolan varies from place to place. For example, for silica fume and GGBS we have more or less the same specification all over the world. But for natural pozzolans it is not the same. Therefore, before using any natural pozzolan in concrete, one should find all the specification of the material.

We didn't use any natural pozzolan to check our mix design process.

2.2.5   Blended Cement

Cement producers sometimes blend above binders with Portland cement to make blended cements. In fact, today because the use of these supplementary cementitious materials is highly recommended, cement factories produce more blended cements.

In cement factories, these binders are mixed with a special percentage with pure Portland cement to make blended cements.

For example, in Europe there is 27 types of cement. Only one type is pure Portland cement and others are blended cements according to table 2.16.

Table 2.16: Types of cements in Europe (EN 197-1)

Cement type	Comments	Derived Types
CEM I	Pure Portland cement	1 type
CEM II	Portland composite cement	19 types
CEM III	Slag cement	3 types
CEM IV	Pozzolanic cement	2 types
CEM V	Composite cement	2 types

In ASTM standard also, there are six main types of blended cements which you can find in table 2.17.

Table 2.17: Types of blended cement in ASTM

Type of cement	Comments
Type IS	Portland slag cement
Type IP	Portland Pozzolanic cement
Type P	Portland pozzulanic cement without high initial strength
Type I(PM)	Modified Portland Pozzolanic cement
Type I(SM)	Modified Portland slag cement
Type S	Portland slag cement to be used with pure Portland cement in concrete

Here locally, like other countries we have blended cements in some factories. But as we don't have good quality fly ash, there is no fly ash blended cement. Regarding high prices of silica fume, cement producers do not blend it with cement. But we can use it separately in concrete. So, here we have slag cements and pozzolanic cements with different percentage of slag and natural pozzolans.

As we use silica fume and GGBS separated from Portland cement for checking our mix design process, it will be more precise. So, we are not going to use blended cement in our process.

2.3   Aggregates

We have two main kinds of aggregates in concrete: coarse aggregates and fine aggregates. In different texts, the boundary of coarse and fine is different. But usually ASTM sieve number 4 or 4.75 mm is the boundary of coarse and fine aggregates. It means that all sizes coarser than 4.75 mm is considered coarse aggregate and all sizes finer than 4.75 mm is considered fine aggregate.

In ASTM standard, we have several tests for aggregates. You can find the most important tests of aggregates in table 2.18. But for concrete mix design, we need sieve analysis of coarse and fine aggregates according to ASTM C136, and specific gravity and water absorption of coarse and fine aggregates according to ASTM C127 and ASTM C128.

Table 2.18: Important tests of aggregates according to ASTM	
Test code	**Comments**
ASTM C40	Organic impurities in fine aggregates
ASTM C117	Materials finer than 75 Micron in aggregates by washing
ASTM C127	Specific gravity and absorption of coarse aggregates
ASTM C128	Specific gravity and absorption of fine aggregates
ASTM C136	Sieve analysis of coarse and fine aggregates
ASTM C227	Potential alkali reactivity of cement aggregate combination
ASTM C289	Potential alkali silica reactivity of aggregates (chemical method)
ASTM C586	Potential alkali reactivity of carbonate rocks

2.3.1   Coarse aggregates

In different countries there are different types of coarse aggregates. For example, we have three types of coarse aggregates to use in common concrete:
- Coarse 12-25: Approximate size of particles between 12 mm to 25 mm
- Coarse 11-19: Approximate size of particles between 11 mm to 19 mm
- Coarse 5-12: Approximate size of particles between 5 mm to 12 mm

Instead of size, we should consider the shape of coarse aggregates for concrete. We have two types of coarse aggregates in this case:

- Natural coarse aggregates: They come from natural minerals of aggregates that are from rivers. These are round shape coarse aggregates.
- Crushed coarse aggregates: They come from huge stones and rocks which are crushed to the size used in concrete.

Most of the time, we should use a mixture of 5-12 coarse aggregates with one of the 11-19 or 12-25 ones. In this case, we will have a good combination of coarse aggregates in our concrete. But sometimes we may not need particles coarser than 12 mm. In that case, we only use 5-12 as the coarse aggregate.

2.3.2   Choosing maximum size of coarse aggregate

In concrete mix design, one of the most important decisions to be made is the maximum size of coarse aggregate in a special concrete. Choosing the best size depends on several factors. In this book, we did many tests for maximum size of aggregates in concrete. You can find our recommendations in table 2.19.

In ACI 211.1 for choosing the maximum aggregate size we recommend:

Maximum size of coarse aggregate should not exceed one fifth of free distance between molds and one third of slab diameter and three forth of free distance between rebars in the structure element.

On the other hand, concrete pump producers have recommended some sizes for aggregates in concrete to get the best pumping pressure. As most of the concretes in structures are pumping concretes, we should consider these recommendations.

We also know that in the same water to cement ratio if we have finer aggregates in our concrete, the compressive strength will be higher. For more than ten years working in a ready mixed concrete plant and producing concrete for so many types of projects and so many structural elements, we can recommend table 2.19 for choosing maximum size of coarse aggregate.

Table 2.19: Recommended maximum size of coarse aggregate		
Structural concrete compressive strength	**Structural element**	**Recommended maximum size of aggregate**
Less than 30Mpa	Foundations	25mm
Less than 30Mpa	Floors, columns, walls	19mm
30Mpa to 45Mpa	Foundations, Floors	19mm
30Mpa to 45Mpa	Walls, columns	12mm
45Mpa to 70Mpa	All kinds of elements	12mm
More than 70Mpa	All kinds of elements	9mm

2.3.3   Fine aggregates

Like coarse aggregates, there are different sizes acceptable for fine aggregates in different countries. For example, we have four sizes of fine aggregates:
- Sand 0-3: Approximate size of particles between 0 to 3 mm.
- Sand 0-5: Approximate size of particles between 0 to 5 mm.
- Sand 0-8: Approximate size of particles between 0 to 8 mm.
- Dune sand: It is the finest type of fine aggregate that comes from deserts.

Like the coarse aggregates in the case of particle shape, we have two kinds of fine aggregates:
- Natural sand: They come from riverbeds. So, they are round shaped fine aggregates.
- Crushed sand: They come from rocks. After crushing and sizing, they can be used for concrete.

There is an important point about fine aggregates that need to be considered. In the sieve analysis test, we have some particles which pass the number 100 ASTM sieve. These are named fillers for concrete. These sizes are very important for the workability of concrete. We have a cement paste and fine mortar in concrete, which effect the workability, segregation and bleeding of concrete. Cement paste is the mixture of cement and water in concrete and fine mortar is the mixture of cement, water and aggregates passed by sieve number 100. So, if in the fine aggregates that we are going to use in the concrete, these sizes are less, we should recover that by using very fine aggregates like dune sand or stone powder.

2.3.4   Aggregates for our mix design check

We will use three types of crushed coarse aggregate to check our mix design process:
- A 12-25 crushed coarse aggregate
- A 11-19 crushed coarse aggregate
- A 5-12 crushed coarse aggregate

Also, we will use two types of fine aggregate to check our mix design process as below:
- A natural 0-8 sand
- A crushed 0-5 sand

We did the sieve analysis test according to ASTM C136. We took the following pictures from the testing process (Fig. 2.2).

Figure 2.2: Scale, sieves and vibrator for sieve analysis test of aggregates

Also, we used ASTM C127 and ASTM C128 for the specific gravity and absorption of our coarse and fine aggregates (Fig. 2.3).

Figure 2.3: Testing process for specific gravity and absorption of aggregates

According to the descriptions in question, you can see the results in the tables from 2.20 to 2.29.

Table 2.20: Sieve analysis for coarse aggregate 12-25				
Sieve size (mm)	Weight of aggregates remaining on sieve (gr)	Weight of aggregates passing the sieve (gr)	Percentage of aggregates remaining on sieve (%)	Percentage of aggregates passing the sieve (%)
25	0	1829	0	100
19	645	1184	35.3	64.7
12.5	879	305	48.1	16.7
9.5	246	59	13.4	3.2
4.75	59	0	3.2	0
Total	1829	-----	100	-----

Table 2.21: Sieve analysis for coarse aggregate 11-19				
Sieve size (mm)	Weight of aggregates remaining on sieve (gr)	Weight of aggregates passing the sieve (gr)	Percentage of aggregates remaining on sieve (%)	Percentage of aggregates passing the sieve (%)
25	0	1908	0	100
19	15	1893	0.8	99.2
12.5	983	910	51.5	47.7
9.5	730	180	38.3	9.4
4.75	133	47	7	2.5
Total	1908	-----	100	-----

Table 2.22: Sieve analysis for coarse aggregate 5-12				
Sieve size (mm)	Weight of aggregates remaining on sieve (gr)	Weight of aggregates passing the sieve (gr)	Percentage of aggregates remaining on sieve (%)	Percentage of aggregates passing the sieve (%)
19	0	1780	0	100
12.5	90	1690	5.1	94.9
9.5	657	1033	36.9	58
4.75	946	87	53.1	4.9
2.36	87	0	4.9	0
Total	1780	-----	100	-----

Table 2.23: Sieve analysis for natural sand 0-8				
Sieve size (mm)	Weight of aggregates remaining on sieve (gr)	Weight of aggregates passing the sieve (gr)	Percentage of aggregates remaining on sieve (%)	Percentage of aggregates passing the sieve (%)
4.75	284	1530	15.7	84.3
2.36	322	1208	17.8	66.6
1.18	298	910	16.4	50.2
0.6	210	700	11.6	38.6
0.3	262	438	14.4	24.1
0.15	330	108	18.2	6
Total	1814	-----	100	-----

Table 2.24: Sieve analysis for crushed sand 0-5				
Sieve size (mm)	Weight of aggregates remaining on sieve (gr)	Weight of aggregates passing the sieve (gr)	Percentage of aggregates remaining on sieve (%)	Percentage of aggregates passing the sieve (%)
4.75	43	1804	2.3	97.7
2.36	603	1201	32.6	65
1.18	483	718	26.2	38.9
0.6	267	451	14.5	24.4
0.3	193	258	10.4	14
0.15	146	112	7.9	6.1
Total	1847	-----	100	-----

Table 2.25: Specific gravity and absorption of Coarse aggregate 12-25	
Specific Gravity (kg/L)	2.795
Water absorption (%)	0.96

Table 2.26: Specific gravity and absorption of Coarse aggregate 11-19	
Specific Gravity (kg/L)	2.791
Water absorption (%)	0.98

Table 2.27: Specific gravity and absorption of Coarse aggregate 5-12	
Specific Gravity (kg/L)	2.785
Water absorption (%)	0.98

Table 2.28: Specific gravity and absorption of Natural sand 0-8	
Specific Gravity (kg/L)	2.699
Water absorption (%)	1.69

Table 2.29: Specific gravity and absorption of Crushed sand 0-5	
Specific Gravity (kg/L)	2.714
Water absorption (%)	1.54

2.4   Water for concrete

It is accepted that the best water for concrete production is drinking water. But as drinking water resources are decreasing every day, it could be better for any industry to use non-drinking water. To control corrosion in concrete structures, it is essential to use defined water for concrete. So, we cannot use any kind of water for concrete production.

Materials Research Forum LLC
https://doi.org/10.21741/9781644900598

Table 2.30: Specifications of non-drinking water for concrete

Type of harmful material	Condition of use	Max acceptable (ppm)
Solid suspended particles	Pre-stressed concrete	1000
	Mild condition[2]	2000
	Severe condition	1000
Total solved materials	Pre-stressed concrete	1000
	Mild condition	2000
	Severe condition	1000
Total chloride content	Pre-stressed concrete	500
	Mild condition	1000
	Severe condition	500
Total sulfate content	Pre-stressed concrete	1000
	Concrete in any condition	1000
Equivalent alkali	For any kind of concrete	600

Standards all over the world define some specification for water that we would like to use for concrete production. For example, table 2.30 shows some definitions for a normal condition that we should use for our local concrete mix designs checking.

2.5   Chemical admixtures

Chemical admixtures for concrete are some chemicals which use less than three percent by weight of binders in concrete. Although their amount of use is less in comparison with other materials, their effects are very high in any case. You can find the most important chemical admixture types in table 2.31.

Table 2.31: Some important chemical admixtures for concrete

Admixtures name	description
Plasticizers (water reducers)	Increase slump, reduce water less than 12% or increase slump and reduce water together
Super-plasticizers (high range water reducers)	Increase slump, reduce water more than 12% or increase slump and reduce water together
Accelerators	Accelerate concrete setting and hardening
Retarders	Retard concrete setting and hardening
Air entraining admixtures	Entrain air bubbles to the concrete to increase freeze thaw resistance
Antifreeze admixtures	Accelerate setting time and increase hydration heat of cement

From the above list, we should consider the specification of plasticizers, super-plasticizers and air entraining admixtures in concrete mix design.

---

[2] These conditions are defined according to Iranian Standard.

2.5.1   *Air entraining admixtures*

When we have freezing and thawing condition in a region, we must use air entraining admixtures to increase concrete resistance to freezing and thawing. These admixtures produce the same size and shape air bubbles in the concrete structure. These air bubbles act as valves to control volume raising of water during freezing process. But on the other hand, these air bubbles decrease concrete compressive strength. ACI 211.1 states that for each percent of air in concrete, the compressive strength will drop three to five percent. With this, we can determine the effect of these admixtures in concrete mix design. But before that, we should understand the amount of air in concretes with air entraining admixtures. For this reason, we can use table 2.32 which is ACI recommendations for concretes in freeze thaw conditions.

Table 2.32: ACI recommendations for concrete in freeze thaw conditions

Max size of coarse aggregates (mm)	Percentage of air in severe condition (%)	Percentage of air in mild condition (%)
9.5	7	6
12.5	7	5
19	6	5
25	6	4

Severe condition: The condition in which concrete is in free cold weather and with high moisture and deicing salts is in the environment like bridge decks
Mild condition: The condition in which concrete is in an environment with low moisture and without deicing salts.

2.5.2   *Plasticizers and super-plasticizers*

Plasticizers and super-plasticizers are the most important chemical admixtures in concrete. There are four main chemical bases of plasticizers and super-plasticizers in the market which you can see in the table 2.33.

Table 2.33: Chemical bases of plasticizers and super-plasticizers

Chemical base	Water reducing rate (%)	Other specifications
Ligno-sulfonate base	8 to 12	Retarding effect, increase entrapped air
Naphthalene sulfonate base	15 to 22	It can accelerate or retard concrete, high range of use
Melamine sulfonate base	12 to 20	High accelerating effect, low entrapped air
Poly Carboxylate ether base	25 to 40	It can accelerate or retard concrete, high range of use, the most powerful

In concrete mix design, we have to know water reducing rate for each dosage of plasticizer or super-plasticizer. For this reason, we need a test to understand these values. In this book, we use a novel test which base is the standard for concrete admixtures[3].

In this standard, we will make a testimonial concrete with defined properties. Then we use concrete plasticizer or super-plasticizer in two ways:
- Concrete with the same slump: In this case, we use a defined dosage of admixture for a defined concrete with the same slump. We then compare this concrete with the testimonial. Of course, the concrete with admixture will need less water for the same slump. Then we will compare percentage of water reduction, compressive strength at one day and twenty-eight days and air entrapped to the concrete by the admixture.
- Concrete with the same water to cement ratio: In this case, we use a concrete using a defined dosage of admixture with the same water to cement ratio with our testimonial concrete. Of course, this concrete will have more slump than the testimonial. Then we will compare the increase of slump after a defined time, compressive strength at twenty-eight days and air entrapped to the concrete by the admixture.

By paying attention to this test method, we have defined a test to obtain the water reduction rate for any plasticizer or super-plasticizer.

2.5.3 The novel test for obtaining water reduction rate of any plasticizer or super-plasticizer using cement mortar

This is a novel test with mini slump cone that is used in DIN EN1015. You can see the dimensions and shape of this cone in figures 2.4 and 2.5.

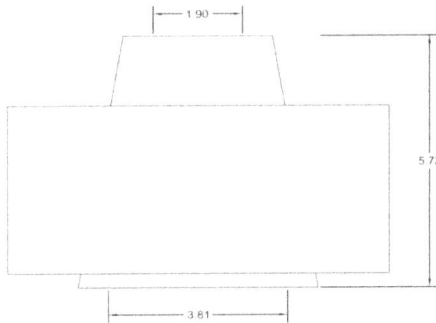

Figure 2.4: Dimensions of mini slump cone

---

[3] ISIRI 2930

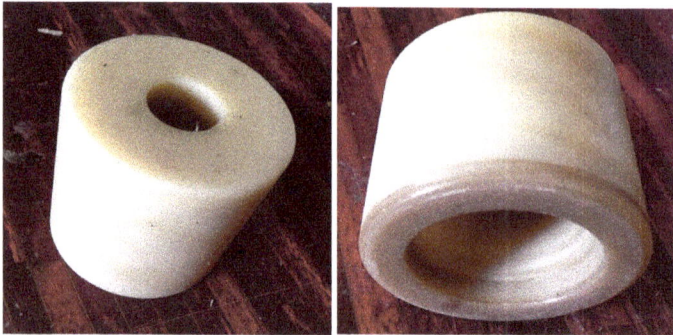

Figure 2.5: Shape of mini slump cone

For this novel test, we use 400 gr of fine quartz sand used for cement compressive strength test according to ASTM C109 and 300 gr of cement. At the first step, we will make a mortar with enough amount of water to achieve a defined flow. Then we will assume at least three dosages for the plasticizer or super-plasticizer that we would like to test, according to the dosage range defined in the product data sheet. For example, if the dosage range is between 0.5 to 1.5, we can use 0.5% and 1.0% and 1.5%. We should make three mortars with these dosages of super-plasticizer. For each mortar, we should decrease the amount of water to achieve the same flow with the testimonial mortar flow. By this technique, we can calculate the rate of water reduction for each dosage.

2.5.4 Super-plasticizer for mix design check

We used a kind of polycarboxylate ether super-plasticizer for our mix design checking with the code BPC-40 and specification shown in table 2.34.

Table 2.34: Specification for admixture BPC-40		
Nominal specification		
Type of Product	Polycarboxylate ether super-plasticizer	
Color	Yellow	
pH	5 to 6	
Specific Gravity	1.08 kg/L	
Solid content	40%	
Tested Specification		
Color	Yellow	Visual
pH	5.1	According to ISIRI 3178-18
Specific gravity	1.083	ISIRI 898
Solid content	38.5%	EN 480-8

We tested this admixture for the water reduction rate according to the novel procedure described above with two kinds of cements that we would like to use.

For the first cement company (A) type II, we used 210 gr of water to achieve 150 mm flow without using any admixture. Then we used 0.5% = 1.5 gr and 1.0% = 3.0 gr and 1.5% = 4.5 gr admixture, and made mortars to obtain the same 150 mm flow. In this process, we used 170 gr, 150 gr and 135 gr of water for the same flow. According to the above descriptions, you can see the results in table 2.35 and figure 2.6.

Table 2.35: Water reduction rate for BPC-40 and company (A) type II cement		
Mortar mix design: Quartz sand: 400gr Company (A) Type II cement: 300gr Water: 210gr Flow on table: 150mm		
Admixture dosage	**Amount of water for 150mm flow**	**Water reduction rate**
0.5%=1.5gr	170 gr	19%
1.0%=3.0gr	150 gr	28.6%
1.5%=4.5gr	135 gr	35.7%

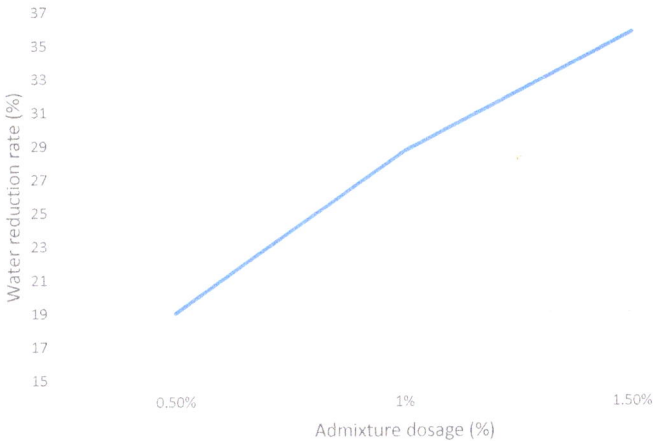

Figure 2.6: Water reduction rate chart for BPC-40 and company (A) cement

As you can see from figure 2.6, we defined two lines for water reduction rate of this admixture and company (A) cement. One line between the dosage of 0.5 and 1 and another line between the dosage of 1 and 1.5.

For this admixture and company (B) type I-525, we repeated this test as well. In this case, we used 225 gr of water to achieve 150 mm flow without using any admixture. Then we used 0.5% = 1.5 gr and 1.0% = 3.0gr and 1.5% = 4.5 gr admixture and make mortars to obtain the same 150 mm flow. In this process, we used 195 gr, 175 gr and 155 gr of water for the same flow. According to the above descriptions, you can see the results in table 2.36 and figure (2.7).

Table 2.36: Water reduction rate for BPC-40 and company (B) type I-525 cement		
Mortar mix design: Quartz sand: 400 gr Company (B) type I-525 cement: 300 gr Water: 225 gr Flow on table: 150mm		
Admixture dosage	**Amount of water for 150 mm flow**	**Water reduction rate**
0.5%=1.5 gr	185 gr	17.8%
1.0%=3.0 gr	165 gr	26.7%
1.5%=4.5 gr	150 gr	33.3%

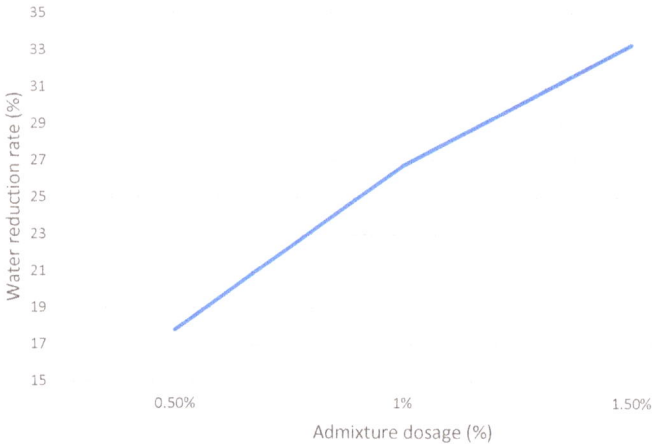

Figure 2.7: Water reduction rate chart for BPC-40 and company (B) cement

2.5.3 Control the water reduction rate for concrete

In this part, we are going to control the obtained water reduction rate for the concrete. We will make a testimonial concrete without any admixture. Then we will use the same mix design with 0.5%, 1.0% and 1.5% of admixture separately. For each concrete, we will calculate the water reduction rate and compare it with the above rates that we obtained from the novel test.

For company (A) cement, you can see the test results in tables 2.36 to 2.39.

Table 2.36: Testimonial concrete for company (A) cement without admixture

Material	For 1 m3	For 22 L
Company (A) type II cement	400 kg	8.8 kg
Aggregate 11-19 (Dry)	430 kg	9.5 kg
Aggregate 5-12 (Dry)	220 kg	4.8 kg
Aggregate Natural sand (Dry)	1150 kg	25.3 kg
Water	245 kg	5.4 kg
Admixture (BPC-40)	0.0 kg	0.0 kg
Slump	180 mm	
Total water absorption of aggregates	26 kg	

Table 2.37: Concrete for company (A) cement with 0.5% of admixture

Material	For 1 m3	For 22 L
Company (A) type II cement	400 kg	8.8 kg
Aggregate 11-19 (Dry)	430 kg	9.5 kg
Aggregate 5-12 (Dry)	220 kg	4.8 kg
Aggregate Natural sand (Dry)	1150 kg	25.3 kg
Water	204 kg	4.5 kg
Admixture (BPC-40)	0.5% = 2.0 kg	44 gr
Slump	185 mm	
Total water absorption of aggregates	26 kg	
Water reduction rate	18.7%	

Table 2.38: Concrete for company (A) cement with 1.0% of admixture

Material	For 1 m3	For 22 L
Company (A)type II cement	400 kg	8.8 kg
Aggregate 11-19 (Dry)	430 kg	9.5 kg
Aggregate 5-12 (Dry)	220 kg	4.8 kg
Aggregate Natural sand (Dry)	1150 kg	25.3 kg
Water	182 kg	4.0 kg
Admixture (BPC-40)	1.0% = 4.0 kg	88 gr
Slump	190 mm	
Total water absorption of aggregates	26 kg	
Water reduction rate	28.7%	

Table 2.39: Concrete for Company (A) cement with 1.5% of admixture

Material	For 1 m3	For 22 L
Company (A) type II cement	400 kg	8.8 kg
Aggregate 11-19 (Dry)	430 kg	9.5 kg
Aggregate 5-12 (Dry)	220 kg	4.8 kg
Aggregate Natural sand (Dry)	1150 kg	25.3 kg
Water	168 kg	3.7 kg
Admixture (BPC-40)	1.5% = 6.0 kg	132 gr
Slump	195 mm	
Total water absorption of aggregates	26 kg	
Water reduction rate	35.1%	

You can see in the above tables that the water reduction rate results are more or less similar to the results obtained from the novel test. So, we can use them for our concrete mix design checking in the future.

For company (B) cement, we also controlled the results by concrete. You can see these results in tables 2.40 to 2.43.

Table 2.40: Testimonial concrete for company (B) cement without admixture

Material	For 1 m^3	For 22 L
Company (B) type I-525 cement	400 kg	8.8 kg
Aggregate 11-19 (Dry)	430 kg	9.5 kg
Aggregate 5-12 (Dry)	220 kg	4.8 kg
Aggregate Natural sand (Dry)	1150 kg	25.3 kg
Water	259 kg	5.7 kg
Admixture (BPC-40)	0.0 kg	0.0 kg
Slump	190 mm	
Total water absorption of aggregates	26 kg	

Table 2.41: Concrete for company (B) cement with 0.5% of admixture

Material	For 1 m^3	For 22 L
Company (B) type I-525 cement	400 kg	8.8 kg
Aggregate 11-19 (Dry)	430 kg	9.5 kg
Aggregate 5-12 (Dry)	220 kg	4.8 kg
Aggregate Natural sand (Dry)	1150 kg	25.3 kg
Water	218 kg	4.8 kg
Admixture (BPC-40)	0.5% = 2.0 kg	44 gr
Slump	180 mm	
Total water absorption of aggregates	26 kg	
Water reduction rate	17.5%	

Table 2.42: Concrete for company (B) cement with 1.0% of admixture

Material	For 1 m3	For 22 L
Company (B) type I-525 cement	400 kg	8.8 kg
Aggregate 11-19 (Dry)	430 kg	9.5 kg
Aggregate 5-12 (Dry)	220 kg	4.8 kg
Aggregate Natural sand (Dry)	1150 kg	25.3 kg
Water	200 kg	4.4 kg
Admixture (BPC-40)	1.0% = 4.0 kg	88 gr
Slump	200 mm	
Total water absorption of aggregates	26 kg	
Water reduction rate	25.3%	

Table 2.43: Concrete for company (B) cement with 1.5% of admixture

Material	For 1 m3	For 22 L
Company (B) type I-525 cement	400 kg	8.8 kg
Aggregate 11-19 (Dry)	430 kg	9.5 kg
Aggregate 5-12 (Dry)	220 kg	4.8 kg
Aggregate Natural sand (Dry)	1150 kg	25.3 kg
Water	182 kg	4.0 kg
Admixture (BPC-40)	1.5% = 6.0 kg	132 gr
Slump	200 mm	
Total water absorption of aggregates	26 kg	
Water reduction rate	33%	

In this case also, you can see from the above tables that the water reduction rate results are similar to the results obtained from the novel test.

Now we have the properties of all constituent materials for the mix design. We can use them for checking the procedure in the future.

3.   Concrete properties

To accept a concrete mix design, we should first test and control it in the laboratory. If the properties of produced concrete with our mix design are compatible with the defined properties, then we can accept our mix design and as we are not going to make any changes in the concrete constituent materials, we can use this mix design industrially.

In this chapter, we will talk about the most important tests that we should do to accept a concrete. These tests are made in fresh concrete and hardened concrete.

3.1   Fresh concrete

In the concrete production process, when cement particles reach the water molecules, the concrete is said to be in the fresh state. This state will continue as long as the concrete remains flexible and pasty and is hence workable. The tests for fresh concrete are done in this state.

3.2   Control the temperature of fresh concrete

All codes and standards introduce a maximum and minimum temperature for fresh concrete. If the concrete temperature even in cold or hot weather condition remain in this range, there is no problem for concreting. However, if the concrete temperature fall outside this temperature the concrete should not be poured.

As you know, minimum fresh concrete temperature may appear in cold seasons of the year and/or in cold climate conditions. In this temperature, the hydration reaction stops or slows down considerably. So, in this temperature, concrete cannot grow its strength and hardening process as well as possible.

On the other hand, maximum temperature of fresh concrete could appear in hot seasons of the year and/or in hot climate conditions. In this temperature, some undesirable reactions of cement will speed up. These reactions may disorder the fundamental reaction of the hydration process. Also, the situation of all reactions in the hydration process will not be separately reliable. So, the concrete cannot gain its strength and durability properties.

Hence it is importance of controlling temperature of fresh concrete. In the local concrete standard that we are going to use, the minimum acceptable temperature of concrete is 5°C and the maximum acceptable fresh concrete temperature is 32°C.

Temperature control of concrete is monitored by a concrete thermometer made especially for concrete. At any time, in the age of fresh concrete, we can test and control the temperature.

In figure 3.1, you can see two types of concrete thermometers.

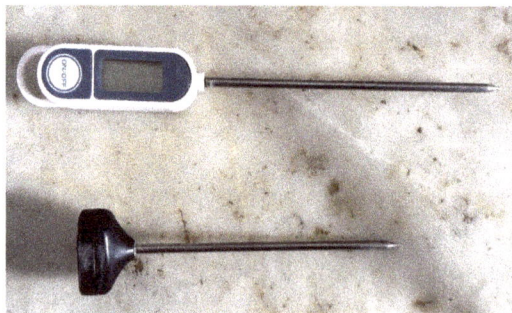

Figure 3.1: Two types of concrete thermometer.

3.2.1 Workability

Workability means the capability of concrete for good compaction and plasticity. Several specifications of concrete constituent materials affect the workability. For example, sieve analysis of total aggregates, amount of cement paste, amount of fine mortar and concrete fluidity affect the workability of concrete. Therefore, aggregates and the amount of cement and fine mortar should be considered in the mix design.

The most important and usual test for fluidity of concrete is the slump test, compatible with ASTM C143. You can see a picture of slump test tools in figure 3.2.

Figure 3.2: Slump test tools.

Before we start the concrete mix design we should consider a target slump for our concrete. Depending on the structural element that we would like to pure, we should select a slump for the best result. There are different suggestions for slump choice in different texts. In this book, with more than seven years of experience with real structures and different elements, and also by paying attention to different recommendations, we can suggest table 3.1 for choosing the slump.

Table 3.1: Slump suggestions for different structural elements	
Type of concrete and structural element	**Suggested slump (mm)**
Foundations and mass structures	100 to 120
Slab floors	150 to 170
Rib and block floors[4]	120 to 170
Columns and walls	170 and more
Thin elements	170 and more
Concretes with Compressive strength of less than 30 MPa	120 to 180
Concretes with compressive strength of 30 to 40 MPa	150 to 180
Concretes with compressive strength more than40 MPa	170 and more

---

[4] This is a kind of common floor in Iran.

To make a concrete with slump of more than 120 mm, one needs to use plasticizers and super-plasticizers.

In some cases, we may use a special concrete like self-compacting concrete (SCC). In this case, instead of a slump test we should do another test to understand the compatibility of concrete with defined targets. For example, for SCC concrete, we do a slump flow test, U box test and L box test. In this book, we will use the slump test for the mix design check.

3.2.2 Entrapped air

There are two kinds of air bubbles in concrete. One is the entrained air which is used to improve concrete durability against freezing and thawing. This kind of air bubbles is supported by using air entraining admixtures. But in all kinds of concrete, there are air bubbles as well. These air bubbles are called entrapped air which are not suitable for concrete at all, but there is no way to eliminate them. If the amount of entrapped air increases, concrete compressive strength will decrease drastically. So, controlling the amount of entrapped air is very important.

To control the amount of air in the concrete, we can use ASTM C173 or ASTM C231 tests. In each case, there is special instruments that we have to use. Explanations of the procedure of estimating concrete entrapped air are in the instrument booklet.
You can see a picture of the instrument for this test in figure 3.3.

Figure 3.3: Instrument for air content test according to ASTM C231

3.3 Hardened concrete

The most important tests of concrete are the ones referring to hardened concrete because the loading capacity and durability of concrete is measurable in the hardened state. The most common age of concrete for testing is twenty-eight days. But we can control it in other ages, less or more than twenty-eight days. For example, we can test hardened concrete at seven days to continue our structure or in ninety days for estimating the behavior of our concrete in the future.

There are several tests for hardened concrete, but here we will talk about the most important ones, which are necessary to control our mix design.

3.3.1 Concrete compressive strength test

The most important test for hardened concrete which is the controller of our mix design method is the compressive strength test. In different countries and with different codes, different types of specimens are used for concrete compressive strength test. For example, we can use 15x15x15 cm cubes and15x30 cm cylinders. This cylinder is the standard test specimen in ACI. It means that if we use cubes or any other types of specimens, we should convert the test result to the cylinder specimen. Of course, it is better to use cylinder specimens especially for higher strengths concrete. But because of the heavy weight of cylinder specimens and its need to capping before test, the laboratory technicians often like to use cube specimens. Hence the need arises to convert the cube compressive strength to a cylinder compressive strength. For this conversion, we can use local concrete code suggestions in table 3.2.

Table 3.2: Converting cube to cylinder compressive strength							
15x15x15cm Cube specimen	25Mpa	30Mpa	35Mpa	40Mpa	45Mpa	50Mpa	55Mpa
coefficient	0.8	0.833	0.857	0.875	0.889	0.9	0.909
15x30cm Cylinder specimen	20Mpa	25Mpa	30Mpa	35Mpa	40Mpa	45Mpa	50Mpa

You can see a picture of cube and cylinder molds in figure 3.4 and cube and cylinder specimens in the figure 3.5.

Figure 3.4: A 15x15x15 cube mold in left and a 15x30 cylinder mold in right

Figure 3.5: A 15x15x15 cube specimen (left) and a 15x30 cylinder specimen (right)

The test procedure of compressive strength for cylinder specimens is according to ASTM C39.

Doing the compressive strength test properly is very important because accuracy in this test can affect directly the test results. Careless testing can cause up to 50% of difference in the test results. The most important errors that can occur in the compressive strength test are:

- Using cube concrete specimens and wrong converting factors.
- The test sample of concrete is not representative of total concrete.
- Errors in the calibration of testing machines.
- Using cube molds that are not rigid enough to give us a soft surface.
- Defects in capping of cylinder specimens.

If the test is correct, the collapse of specimens should be in the right manner. For cubes and cylinders, the failed specimen should be like two separate cones. You can see a picture of a good capped cylinder specimen in figure (3.6) and a 300 tons test machine that we use in figure 3.7.

Figure 3.6: A good capped cylinder specimen

Figure 3.7: A 300 tons concrete test machine

3.3.2   *Permeability of concrete*

Instead of compressive strength, another important specification of concrete is its durability. Calculating the durability of concrete is not simple. But there are some techniques to estimate it. Most of these techniques are based on the permeability of concrete.

The lower the permeability of concrete, the higher the durability. With low permeability, the corrosive ions and also water cannot leak to the concrete structural element. So, corrosion will be impossible. To decrease the permeability of concrete as little as possible, we should consider the following actions:

- Lower the water to cement ratio
- Use of supplementary cementitious materials like silica fume, fly ash and GGBS.
- Design of concrete from highest particle size to the lowest particle size in the best form to achieve a concrete with good particle size distribution.
- Cure concrete well (control the humidity and temperature during the time)
- Using concretes with higher compressive strength.

For testing the permeability of concrete specimens we have two tests. One is ASTM C642 and the other test is the rapid chloride permeability test[5] that shows the permeability of chloride ions by a rapid electrical method.

4.   The novel step by step procedure for the concrete mix design

In this section, we will talk about the step by step procedure for the concrete mix design. First, we will briefly talk about this procedure and then we will test it with different kinds of concrete (different compressive strength classes of concrete). We will test the concrete samples in the laboratory and then test them in a batching plant for defined projects.

4.1   Goals of the concrete mix design

In a concrete mix design, our goals are to achieve three properties of concrete: Compressive strength, fluidity and durability. On the other hand, we should consider economic terms. So, we should try to achieve a mix design with qualified properties and economic advantages.

4.2   Step (1): Specify standard deviation

Standard deviation is a digit that should exceed the structural compressive strength to achieve a mix design compressive strength. Because of so many issues that may occur during the concrete production, we should design a concrete with a compressive strength higher than the structural compressive strength to confirm the structural properties in the elements.

For standard deviation, we will use two procedures depending on if we have any previous data from our batching plant or not.

4.2.1   First method for standard deviation: Calculation from previous information

This method is used when we have some data of compressive strength in the same plant and project beforehand. Two projects are in the same conditions if the following specifications are identifiable:
- The concrete constituent materials in the two projects should commonly be the same from a technical point of view.
- Quality control and supervision of the two projects should be the same.
- The difference in structural concrete compressive strength in the two projects should be less than 5 Mpa.

In this case, we can use the first method to specify standard deviation with the use of this equation:

---
[5] RCPT or ASTM C1202

$$S = \sqrt{\frac{\sum(x - m)^2}{n - 1}}$$
4.1

In this equation S is the standard deviation, X is the compressive strength of a specimen, m is the mean value of compressive strengths of the specimens and n is the number of specimens.

We can use equation 4.1, if we have at least thirty specimens. If we have less, we should multiply a coefficient to the above standard deviation. The coefficient is specify by this equation:

$$R = 0.75 + \sqrt{\frac{2}{n}}$$
4.2

In the above equation, R is the coefficient and n, is the number of specimens.
It is possible to have two series of tests and each one has less than thirty specimens. In this case, we should use this equation:

$$S = \sqrt{\frac{(n_1 - 1)s_1^2 + (n_2 - 1)s_2^2}{n_1 + n_2 - 2}}$$
4.3

In this equation, S is the final standard deviation, S_1 and S_2 are the standard deviation of each series of tests and n_1 and n_2 are the numbers of specimens in each series.

It is very important that in any case we cannot define the standard deviation less than 2.5 MPa.

4.2.2 Second method for standard deviation

We should use this second method when we don't have any information from similar projects. In this case, the standard deviation should be more than the first method because we don't have any information beforehand. So, we need a larger safety factor.

For standard deviation in this method, we should first specify the status of our plant from a quality control point of view from table 4.1.

As you see in the table 4.1, status C is not a good and acceptable quality system for concrete production. So, for high quality concrete we should use status A or at least B.

After we specify the status of our production plant from table 4.1, then we should use table 4.2 for the standard deviation.

Materials Research Forum LLC
https://doi.org/10.21741/9781644900598

Table 4.1: Status of a plant from quality control point of view

Conditions of production and QC	Status A	Status B	Status C
Measuring of cement	By weight	By weight	By volume
Measuring of aggregates	By weight	By weight	By volume
Sieve analysis of aggregates	Controlled	Controlled	Not controlled
Moisture of aggregates	Controlled	Controlled	Not controlled
Surveillance of production	Very good	good	weak
Lab instruments	Full available	Available	Not available
Experiments	Continuous	Sometimes	Sometimes
Expert product manager	Available	Available	Not available

Table 4.2: Standard deviation defined with the status of plant and structural compressive strength

Status of production plant	f'c between 20 – 25 MPa	f'c between 25 – 30 MPa	f'c between 30 – 35 MPa	f'c more than 35MPa
A	3 MPa	3.5 MPa	4 MPa	4.5 MPa
B	4 MPa	4.5 MPa	5 MPa	-----
C	5 MPa	-----	-----	-----

You can see in table 4.2 that status C can only be used for structural compressive strength less than 25 MPa and for concretes structural compressive strength more than 35 MPa. You should only use a status A plant because for high strength concrete, quality control is more important than for normal concrete.

4.3 Step (2): Specify mix design compressive strength

Mix design compressive strength is a strength value which we are going to use for our mix design. We should use the following equation to find the mix design compressive strength:

$$f_{cm} = f'_c + 1.34\,s + 1.5 \qquad\qquad 4.4$$

In the above equation, f_{cm} is the mix design compressive strength, f'c is the structural compressive strength and s is the standard deviation which was specified earlier.

4.4 Step (3): Specify percentage of each aggregate in the concrete

This step of mix design is very important. It needs special attention and accuracy. On the other hand, the users experience in working with concrete is very important.

First we should specify the maximum size of the coarse aggregate for our concrete. For several uses, we need different maximum sizes of coarse aggregate. For example, maximum size of coarse aggregate in a mass foundation compared to a narrow column with congested rebars should be different.

In section 2.3.2 we talked about specification of maximum size of coarse aggregate. So, you can use that section especially table 2-19 to find the maximum size of coarse aggregate for any kind of concrete.

After finding the maximum size of coarse aggregate, we can use the following equation which is the modified Voler-Thompson equation to specify the percentage of each kind of aggregate:

$$P = \frac{100\%}{1 - \left(\frac{0.075}{D}\right)^n} \times \left[\left(\frac{d}{D}\right)^n - \left(\frac{0.075}{D}\right)^n\right]$$

                   4.5

In the above equation, P is the percentage passed from sieve d, and D is the maximum size of coarse aggregate in mm, and n is a digit between 0.1 to 0.7 that specifies the coarseness or fine texture of the concrete. In fact, when we use smaller values for n our concrete texture will be finer and when we use bigger values for n, our concrete texture will be coarser. To find best values of n, you can use table 4.3.

You should use two values for n, one is the upper limit and the other one is the lower limit.

Table 4.3: Values of n for different kinds of concrete		
Type of concrete	**Value of n related to minimum curve**	**Value of n related to maximum curve**
Fine SCC concrete	0.2	0.1
Coarse SCC concrete	0.25	0.15
High slump concrete (more than 170 mm)	0.3	0.2
Pumpable concrete with slump between 140 to 170 mm	0.35	0.25
Not pumping concrete with slump between 140 to 170 mm	0.4	0.3
Not pumping concrete with slump less than 140 mm	0.45	0.35
Not pumping concrete for huge elements	0.5	0.4

4.5 *Step (4): Specify fineness module of total aggregates*

The fineness module of the total aggregates is defined as the sum of cumulative percentage remaining with sieves 37.5 mm, 19 mm, 9.5 mm, 4.75 mm, 2.36 mm, 1,18 mm, 0.6 mm, 0.3 mm and 0.15 mm divided by 100. We will use this value in the next steps to find the concrete water demand.

4.6   Step (5): Specify water to binder ratio

This step and the next step are the most important steps in the concrete mix design. So the accuracy in these steps will give us a concrete with exact defined specification. On the other hand, finding the exact value of water to binder ratio for a defined concrete is quite difficult.

To specify water to binder ratio, we first assume that we are going to make our concrete with only Portland cement without other binders. Then we get the information from the cement mortars compressive strength according to ASTM C109. With this information we will define two kinds of cements:
-   Cement I-525: Compressive strength of this cements mortar at 28 days according to ASTM C109 should be more than 525kg/cm2.
-   Cement I-425: Compressive strength of this cements mortar at 28 days according to ASTM C109 should be more than 425kg/cm2.

These cements are two types of ASTM type I standards. But for so many projects, it is possible for us to use another kind of cements for example ASTM type II or V. We can have ASTM C109 tests results for other kinds of cements.

In the next section, we will have three tables of w/b ratio for I-525 cement and three tables of w/b ratio for I-425 cement. Each table is derived from one mix design method. The first table is from a German mix design method, the second table is from an ACI concrete mix design method and the third table is from the mix design method for HPC concrete.

To obtain w/b ratio for our concrete mix design, first we should derive digits for w/b ratio from tables of I-525 and I-425 cement. Then by paying attention to the compressive strength of our cement according to ASTM C109 test, we should find the digits for our cement by linear correlation. Finally, we should choose a w/b ratio from these digits.

Again, you should pay attention that we are talking only about Portland cement as our total binder. It is possible to later add other binders, but for now we only assume that we are going to make our concrete with pure Portland cement.

For I-525 cement we will have table 4.4 and for I-425 cement we will have table 4.5 for w/b ratio derived from German concrete mix design method (method one). This tables contain only concretes with f_{cm} less than 55MPa.

Table 4.4: w/b ratio for concrete made with I-525 cement (Method one)

f_{cm}	w/b for crushed aggregates	w/b for natural aggregates
25	0.69	0.66
30	0.65	0.61
35	0.6	0.57
40	0.55	0.53
45	0.52	0.48
50	0.48	0.44
55	0.44	0.39

Table 4.5: w/b ratio for concrete made with I-425 cement (Method one)

f_{cm}	w/b for crushed aggregates	w/b for natural aggregates
25	0.65	0.62
30	0.59	0.56
35	0.54	0.51
40	0.5	0.47
45	0.46	0.42
50	0.41	0.36
55	0.36	0.34

Also, we will have table 4.6 for w/b ratio from ACI concrete mix design method (method two). This table contains only concrete with f_{cm} less than 45 MPa. In this method, we assume that we are going to use crushed gravels with natural ground sand. In this method, there is no definition for the compressive strength of cement so this shows that this method is less accurate than the previous one.

Table 4.6: w/b ratio from method two

f_{cm}	w/b
25	0.69
30	0.61
35	0.53
40	0.47
45	0.43

On the other hand, we will have w/b ratio from table 4.7 for cement I-525 and table 4.8 for cement I-425 derived from the HPC concrete mix design method (method three). For this method, we will have w/b for f_{cm} between 45 to 80 MPa. In this case, we also assume that we are going to use crushed gravels with natural ground sand.

Table (4.7): w/b ratio for I-525 cement from method three

f_{cm}	w/b
45	0.42
50	0.4
55	0.39
60	0.37
65	0.36
70	0.34
75	0.33
80	0.32
85	0.3

Table 4.8: w/b ratio for I-425 cement from method three

f_{cm}	w/b
45	0.38
50	0.36
55	0.35
60	0.34
65	0.32
70	0.31
75	0.29
80	0.28
85	0.26

As described before, for any kind of cement we should choose a digit for w/b ratio by using data from the above tables. This would be one of the most important parts of our mix design.

4.7   Step (6): Specify free water in concrete

As mentioned before, specifying water to binder ratio and free water in concrete are the most important steps in a concrete mix design. In fact, by following these steps, we have gone most of the way to our final concrete mix design.

Free water means total amount of water in concrete. This is the sum of the surplus or the lack of water in the aggregates and the water we introduce to the concrete in the production process. This free water depends on several factors, such as the sieve analysis of aggregates, target slump and use of plasticizers and super plasticizers.

To specify amount of free water in concrete, you should first pay attention to tables 4.9 and 4.10. In these tables you can find that the amount of free water depends on the amount of cement and the fineness module of total aggregates. Table 4.9 is for targeted slump of about 90 mm and table 4.10 is for targeted slump of about 150 mm. For more or less slump, we can use a linear correlation to find the value. For the amounts of cement

Materials Research Forum LLC
https://doi.org/10.21741/9781644900598

between the values of table, too, you can use linear correlation. For other binders, you can assume that total binder is pure Portland cement.

Table 4.9: Amount of water for about 90 mm slump (F: fineness module of total aggregates)

Amount of cement	F= 4.0	F= 4.1	F= 4.2	F= 4.3	F= 4.4	F= 4.5	F= 4.6	F= 4.7	F= 4.8	F= 4.9	F= 5.0	F= 5.1	F= 5.2	F= 5.3	F= 5.4	F= 5.5
300kg	214	210	206	203	199	196	193	189	186	183	180	177	175	172	169	167
325kg	218	214	210	207	203	200	197	193	190	187	184	181	179	176	173	171
350kg	222	218	214	211	207	204	201	197	194	191	188	185	183	180	177	175
375kg	226	222	218	215	211	208	205	201	198	195	192	189	187	184	181	179
400kg	230	226	222	219	215	212	209	205	202	199	196	193	191	188	185	183
425kg	234	230	226	223	219	216	213	209	206	203	200	197	195	192	189	187
450kg	238	234	230	227	223	220	217	213	210	207	204	201	199	196	193	191
475kg	242	238	234	231	227	224	221	217	214	211	208	205	203	200	197	195
500kg	246	242	238	235	231	228	225	221	218	215	212	209	207	204	201	199

Table 4.10: Amount of water for about 150 mm slump (F: fineness module of total aggregates)

Amount of cement	F= 4.0	F= 4.1	F= 4.2	F= 4.3	F= 4.4	F= 4.5	F= 4.6	F= 4.7	F= 4.8	F= 4.9	F= 5.0	F= 5.1	F= 5.2	F= 5.3	F= 5.4	F= 5.5
300kg	232	228	224	220	216	213	209	206	202	199	196	193	190	187	184	181
325kg	236	232	228	224	220	217	213	210	206	203	200	197	194	191	188	185
350kg	240	236	232	228	224	221	217	214	210	207	204	201	198	195	192	189
375kg	244	240	236	232	228	225	221	218	214	211	208	205	202	199	196	193
400kg	248	244	240	236	232	229	225	222	218	215	212	209	206	203	200	197
425kg	252	248	244	240	236	233	229	226	222	219	216	213	210	207	204	201
450kg	256	252	248	244	240	237	233	230	226	223	220	217	214	211	208	205
475kg	260	256	252	248	244	241	237	234	230	227	224	221	218	215	212	209
500kg	264	260	256	252	248	245	241	238	234	231	228	225	222	219	216	213

4.7.1 Using plasticizers and super-plasticizers and their effect in free water

As mentioned before in 2.5.3 to 2.5.5, we can obtain plasticizers and super-plasticizers water reduction rate with the novel test method. In this case, when we would like to use a kind of super-plasticizer in our concrete, we should first do the novel test and obtain water reducing rate of that admixture. After that, we should decide to use a dosage of the super-plasticizer for our concrete with specified characteristics. To do that, we should consider the economy, chemical base of super-plasticizer and the target slump of our concrete, too. Instead of these, we can do a marsh cone test to specify saturation point of our admixture. It will help us to find a good dosage of admixture for our concrete.

The saturation point is the amount of plasticizer or super-plasticizer which is the maximum dosage. Using more does not have an effect on the plasticizing behavior of the admixture. Of course, saturation point of admixtures with different chemical bases are different. So, it can give us an idea to decide on the dosage of our super-plasticizer by paying attention to the economical comparisons. For marsh cone test, you can see the appendix 2.

Another way to make a good decision about the best dosage of super-plasticizer is paying attention to the recommendations of super-plasticizer manufacturer. They specify a dosage range for their product and they also have some recommendations on which dosage is best for which concrete.

For example, for the super-plasticizer that we are going to use for our mix design check, (BPC-40) you can see appendix 3 for Material Date Sheet of this admixture.

Finally, we will specify a dosage of admixture and from the figure of water reduction rate, we can specify a percentage of water reduction. After that we should decrease the amount of free water by considering the percentage of water reduction rate.

4.8   Step (7): Specify the amount of cement and other binders

To obtain total binder we can use equation 4.6:

$$b = \frac{w}{(w/b)} \qquad\qquad 4.6$$

In this equation, b is the total amount of binder, w is the amount of final reduced free water and w/b is the water to binder ratio from step (5).

In some kinds of concrete, it is possible to use only Portland cement. Of course, in these concretes, b is the amount of Portland cement. But when we would like to use other binders with Portland cement for our concrete[6], we should specify the amount of Portland cement and other binders separately. To do that, we should follow this procedure:

- Specify b from equation 4.6
- After that from table 2.12 and by bringing into account the terms of the project, we decide the percentage of any binder used instead of cement. For example, we decide to use 30 percent of GGBS.
- Using other binders as a percentage of total b instead of using pure Portland cement increases the durability, but on the other hand it decreases the strength of the concrete especially in its early days. Therefore, we have to specify a percentage to increase the amount of b by considering the type of pozzolan[7]. To do so, we use table 4.11.

---

[6] In concrete industry, most of the time we should use other binders for best results in all specification of concrete especially durability.
[7] In the European standard k is a coefficient for this reason.

Table 4.11: Coefficient of increasing b correlated with type of pozzolan	
Type of binder	**Percent of increasing b**
Silica Fume	0%
Fly ash	10%
GGBS	15%
Natural Pozzolans	10% to 15%

- By first using the coefficient from table 4.11, we should increase the amount of b.
- The amount of pozzolan will be obtained by multiplying the percentage from table 2.12 in the increased b.
- The amount of cement will be obtained by subtracting the amount of pozzolan from increased b.

4.9   Step (8): Specify the total volume of aggregates in concrete

To specify total volume of aggregates in concrete, we use equation 4.7:

$$V = 1000 - \left(\frac{C}{d_c}\right) - \left(\frac{P}{d_p}\right) - w - V_a - \left(\frac{S}{d_s}\right) \qquad 4.7$$

In the equation above, V is the total volume of aggregates in liter, C is the weight of cement in kg, dc is the specific gravity of cement in kg/L, P is the weight of other binders in kg, d_p is the specific gravity of other binders in kg/L, w is the weight of water which is equal to its volume in liter, S is the weight of chemical admixture in kg, d_s is the specific gravity of chemical admixture and finally V_a is the volume of air in 1 m^3 of concrete in liters.

To estimate V_a in a concrete, we can use tables 4.12 and 4.13. Table 4.12 is the correlation between entrapped air in concrete and maximum size of coarse aggregate, and table 4.13 is the correlation between chemical base of super-plasticizer and entrapped air in concrete. By using these two tables together, we can estimate the amount of air in concrete.

Table 4.12: correlation between entrapped air and max size of aggregate			
Max size of aggregates (mm)	12.5	19	25
Percentage of entrapped air (%)	0.8 to 1.5	0.5 to 1.2	0.5 to 1.0

Table 4.13: Correlation between entrapped air and chemical base of admixture				
Chemical base of admixture	Ligno sulfonate	Naphthalene sulfonate	Melamine sulfonate	Poly Carboxylate ether
Percentage of entrapped air (%)	1.1 to 2.0	0.9 to 1.6	0.5 to 1.2	1.0 to 1.8

4.10   Step (9): Calculating the weight of aggregates in saturated surface dry (SSD) condition

By multiplying the percentage of each of the aggregates from step (3) in total volume of aggregates from step (8), we can calculate the volume of each kind of aggregate in our concrete. Then by multiplying the saturated surface dry specific gravity of each kind of aggregate in each volume, we can calculate the weight of each kind of aggregate in the saturated surface dry condition.

4.11   Step (10): Calculating the real weight of aggregates and water in concrete

Aggregates in concrete are always in two conditions. Some of the aggregates are dry. It means that these aggregates have less water than the saturated surface dry condition. So, they will absorb some water from the free water in the concrete. Some other aggregates are wet. It means that these aggregates have more water than the saturated surface dry condition. As a result, they will release some water to the free water in the concrete.

For dry aggregates, we must calculate the amount of water they absorb and add it to the free water in the concrete. Also, we must subtract this weight from the weight of dry aggregate. For wet aggregates we must calculate the amount of water they release and subtract that from the free water in the concrete. Also, we must add this weight to the weight of wet aggregate.

4.12   Making trial mix and control the specification for fresh and hardened concrete

Now our mix design is complete. First, we start our trials in the laboratory. To obtain the compatibility of our mix design with defined concrete, we centralize on two tests. One in the fresh state which is slump test and one in the hardened state which is compressive strength test.

After making concrete with specified mix design accurately, first we test the slump. If the slump of concrete is less than defined, it is better to use a little more of the super-plasticizer. If the slump is more than defined, we decrease the amount of water. After these adjustments, our fresh concrete specifications will be good. Then we make specimens of this concrete and test the compressive strength in seven and twenty-eight days. If it is satisfactory, then we can use this mix design in our batching plant. If the compressive strength is less than defined, we modify our mix design by modifying w/b. If the compressive strength is more than defined, we can decrease the amount of cement. For concrete mix design we suggest to use table 4.14.

Table 4.14: The mix design table				
Batching Plant Name:				
Structural Compressive Strength (MPa):				
Standard Deviation (MPa):				
Mix design Compressive Strength (MPa):				
Aggregate Type	Producer	Percentage of Usage (%)	Water Absorption (%)	Water Content (%)
12-25 mm				
5-12 mm				
Sand No.1				
Sand No.2				
Fineness Module of Total Aggregates:				
Type and Producer of Cement:				
Compressive Strength of Cement (kg/cm^2):				
Other Binders Type:				
Percentage of Usage for Other Binder (%):				
w/b:		Free Water (L):		
Structural Element:				
Target Slump:				
Type of Super-Plasticizer:				
Dosage of Super-Plasticizer (%):		Water Reducing Rate (%):		

Final Mix		
Constituent Material	**Weight For 1m^3**	**Weight For 1 Batch**
Aggregate 12-25 mm		
Aggregate 5-12 mm		
Sand No.1		
Sand No.2		
Portland Cement		
Other Binder		
Water		
Super-Plasticizer		
Other		
Total		

Other Descriptions:

5. Implementing mix design for a sample ready mixed plant

In this chapter, we are going to use our mix design method in a real ready mixed plant for concrete production.

First, we will specify all kinds of concrete that we are going to use in this plant. Then we will calculate the mix design for each kind of concrete by using our novel method. Finally, we will check our mix design in the lab and in the batching plant.

5.1   Specification of the ready mixed plant

Our ready mixed plant is a normal plant in the city of Esfahan, Iran. The list of machinery and instruments of this plant are as below:
-    One pan mixer concrete LIEBHERR batching with the capacity of 50 m^3/h.
-    One pan mixer concrete LIEBHERR batching with the capacity of 30 m^3/h.
-    Number of truck mixers with the capacity of 7 m^3: 6
-    Number of truck mixers with the capacity of 9 m^3: 6
-    One concrete stationary SCHWING pump.
-    One concrete boom SCHWING 42 m pump.
-    One concrete boom REICH 38 m pump.
-    One concrete boom ELBA 26 m pump.
-    Laboratory with all kind of instrument for concrete testing according to Iranian 6037 standard[8].

Figures 5.1 to 5.3 show some of the mentioned machinery.

Figure 5.1: Pan mixer concrete LIEBHERR batching with the capacity of 50 m^3/h and a 9 m^3 truck mixer of the plant.

---

[8] Iranian standard for ready mixed concrete.

Figure 5.2: Pan mixer concrete LIEBHERR batching with the capacity of 30 m^3/h and a 7 m^3 truck mixer of the plant.

Figure 5.3: One concrete boom REICH 38 m pump

5.2   *Different kinds of concrete in the ready mixed plant*

In this part, we are going to specify different kinds of concrete that we are going to use in the ready mixed plant. You can see that in table 5.1.

Type of Concrete	Specifications for production	Uses
C25	With Portland cement + coarse gravel	All structural elements
C30.1	with Portland cement + coarse gravel	All structural elements
C30.2	with Portland cement + fine gravel	Congested elements
C30.3	With Portland cement+ GGBS +coarse gravel	All structural elements
C30.4	With Portland cement+ GGBS+ fine gravel	Congested elements
C35.1	With Portland cement + coarse gravel	All structural elements
C35.2	With Portland cement + fine gravel	Congested elements
C35.3	With Portland cement + GGBS +coarse gravel	All structural elements
C35.4	With Portland cement + GGBS + fine gravel	Congested elements
C40.1	With Portland cement + Coarse gravel	All structural elements
C40.2	With Portland cement + fine gravel	Congested elements
C40.3	With Portland cement + GGBS +coarse gravel	All structural elements
C40.4	With Portland cement + GGBS +fine gravel	Congested elements
C45.1	With Portland cement + fine gravel	All structural elements
C45.2	With Portland cement + GGBS +fine gravel	All structural elements
C50.1	With Portland cement + GGBS+ fine gravel	All structural elements
C50.2	With Portland cement + GGBS + Silica fume + fine gravel	All structural elements
C60	With Portland cement + GGBS + Silica fume + fine gravel	All structural elements
C70	With Portland cement + GGBS + Silica fume + fine gravel	All structural elements

Table 5.1: All kinds of concrete in the ready mixed plant

As you can see in table 5.1, we have 19 different kinds of concrete in this plant.
Now we calculate a concrete mix design for each kind with our novel method and then check them.

5.3 Calculations for concrete mix design
Using the step by step procedure we calculate the concrete mix design for each kind of concrete.

5.3.1 Step (1): Specify standard deviation

In this plant, we have data from the previous concrete production. The products were C25 concrete. Now we are going to use this data for the calculation of the standard deviation. You can see the data for the previous C25 concrete in table 5.2.

Table 5.2: Compressive strength for C25 concrete in our plant in (kg/cm^2)

261	308	295	279	311	251	262	265	297	309
322	253	270	307	311	263	279	298	255	316
257	299	326	254	319	256	261	298	316	300
255	261	279	256	316	259	277	283	295	277

As we have forty specimens, we can use the equation (4.1) as below:

$$S = \sqrt{\frac{\sum (x - m)^2}{n - 1}}$$

The calculations with an excel file done as shown in figure 5.4:

	Data										
	261	308	295	279	311	251	262	265	297	309	
	322	253	270	307	311	263	279	298	255	316	
	257	299	326	254	319	256	261	298	316	300	
	255	261	279	256	316	259	277	283	295	277	
Sum	1095	1121	1170	1096	1257	1029	1079	1144	1163	1202	11356
										m	284
	(x-m)2										
	524	581	123	24	734	1082	480	357	172	630	
	1452	955	193	534	734	437	24	199	835	1030	
	724	228	1772	894	1232	778	524	199	1030	259	
	835	524	24	778	1030	620	48	1	123	48	
Sum	3535	2288	2113	2230	3731	2918	1076	756	2160	1967	22774
										S	24

Figure 5.4: Excel calculations for standard deviation.

As mentioned before, we cannot use standard deviation less than 25 kg/cm^2 or 2.5 MPa. So, we use 2.5 MPa for this ready mixed plant (table 5.3).

Table 5.3: Standard deviation for our ready mixed plant	
Standard deviation for all concretes	2.5 MPa

5.3.2   Step (2): Specify mix design compressive strength

For this step, we use the equation (4.4) as below:

$$f_{cm} = f_c' + 1.34\,s + 1.5$$

The calculations of this equation are in table 5.4.

Table 5.4: Calculations for mix design compressive strength

f'c (Mpa)	S (Mpa)	Fcm (Mpa)
25	2.5	29.85 = 30
30	2.5	34.85 = 35
35	2.5	39.85 = 40
40	2.5	44.85 = 45
45	2.5	49.85 = 50
50	2.5	54.85 = 55
60	2.5	64.85 = 65
70	2.5	74.85 = 75

5.3.3   Step (3): Specify percent of each aggregate in the concrete

As you can see in table 2.19, it is suggested to use 25 mm as the max size of coarse aggregate for concrete less than C30 and foundations and for other structural elements 19 mm. For C30 to C45 we can make two kinds of concrete, one with the max size of coarse aggregate 19 mm and the other 12 mm. For above C45 concretes, we use 12 mm coarse aggregate as the max size. Table 5.5 shows the mixture of all aggregates for each concrete.

Table 5.5: Max size of aggregate for each concrete and the mixture

Type of concrete	Max size of coarse aggregates	Mixture of aggregates
C25.1	25 mm	No.1: Coarse 12-25, Coarse 5-12, Sand 0-8, Sand 0-5
C25.2	19 mm	No.2: Coarse 11-19, Coarse 5-12, Sand 0-8, Sand 0-5
C30.1	19 mm	No.2: Coarse 11-19, Coarse 5-12, Sand 0-8, Sand 0-5
C30.2	12 mm	No.3: Coarse 5-12, Sand 0-8, Sand 0-5
C30.3	19 mm	No.2: Coarse 11-19, Coarse 5-12, Sand 0-8, Sand 0-5
C30.4	12 mm	No.3: Coarse 5-12, Sand 0-8, Sand 0-5
C35.1	19 mm	No.2: Coarse 11-19, Coarse 5-12, Sand 0-8, Sand 0-5
C35.2	12 mm	No.3: Coarse 5-12, Sand 0-8, Sand 0-5
C35.3	19 mm	No.2: Coarse 11-19, Coarse 5-12, Sand 0-8, Sand 0-5
C35.4	12 mm	No.3: Coarse 5-12, Sand 0-8, Sand 0-5
C40.1	19 mm	No.2: Coarse 11-19, Coarse 5-12, Sand 0-8, Sand 0-5
C40.2	12 mm	No.3: Coarse 5-12, Sand 0-8, Sand 0-5
C40.3	19 mm	No.2: Coarse 11-19, Coarse 5-12, Sand 0-8, Sand 0-5
C40.4	12 mm	No.3: Coarse 5-12, Sand 0-8, Sand 0-5
C45.1	12 mm	No.3: Coarse 5-12, Sand 0-8, Sand 0-5
C45.2	12 mm	No.3: Coarse 5-12, Sand 0-8, Sand 0-5
C50.1	12 mm	No.3: Coarse 5-12, Sand 0-8, Sand 0-5
C50.2	12 mm	No.3: Coarse 5-12, Sand 0-8, Sand 0-5
C60	12 mm	No.3: Coarse 5-12, Sand 0-8, Sand 0-5
C70	12 mm	No.3: Coarse 5-12, Sand 0-8, Sand 0-5

The next problem to be solved is the "n" value. Table 4.3 shows us that we have values of "n" between 0.25 to 0.35 for pumping concrete with slump 140 to 170 mm. This concrete is very good for ready mixed plants. So, we use these values of "n" for all kinds of concrete in this plant.

For all types of concrete according to table 5.5, we will make an excel file and calculate the percentage of each kind of aggregates. For No.1 mixture with the max size of coarse aggregate 25 mm we have figure 5.5.

			Max size of agg (mm)	Value for n					Max Size of agg (mm)	Value for n
			25	0.25					25	0.35

0.15 mm	0.3 mm	0.6 mm	1.18 mm	2.36 mm	4.75 mm	9.5 mm	12.5 mm	19 mm	25 mm	Sieve size
4.1	9.4	16.1	24.5	35.3	49.3	66.9	75.2	89.5	100.0	n1 value
5.8	12.7	20.8	30.3	41.8	55.6	71.9	79.2	91.3	100.0	n2 value

Percent (%)	Type	Aggregate
27	Crushed	Coarse Gravel
15	Crushed	Fina Gravel
29	Natural	Sand No.1
29	Crushed	Sand No.2
100	*****	Total

0.15 mm	0.3 mm	0.6 mm	1.18 mm	2.36 mm	4.75 mm	9.5 mm	12.5 mm	19 mm	25 mm	Aggregate
0	0	0	0	0	0.0	0.9	4.5	17.5	27.0	Coarse Gravel
0	0	0	0	0.0	0.7	8.7	14.2	15.0	15	Fine Gravel
1.7	7.0	11.2	14.5	19.3	24.5	29	29	29	29	Sand No.1
1.8	4.1	7.1	11.3	18.9	28.3	29	29	29	29	Sand No.2
3.5	11.1	18.3	25.8	38.2	53.5	67.6	76.7	90.5	100.0	Total

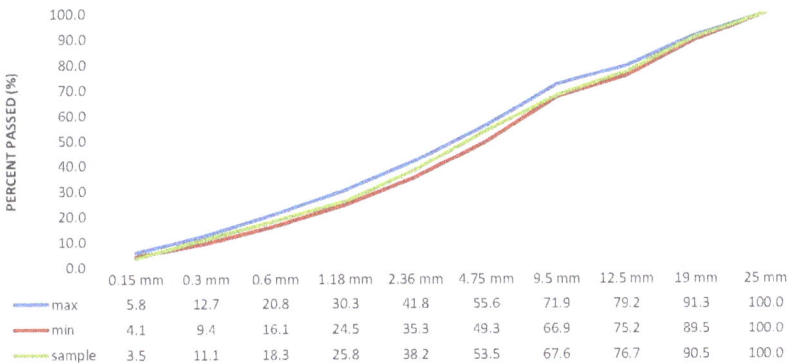

Analysis Chart

	0.15 mm	0.3 mm	0.6 mm	1.18 mm	2.36 mm	4.75 mm	9.5 mm	12.5 mm	19 mm	25 mm
max	5.8	12.7	20.8	30.3	41.8	55.6	71.9	79.2	91.3	100.0
min	4.1	9.4	16.1	24.5	35.3	49.3	66.9	75.2	89.5	100.0
sample	3.5	11.1	18.3	25.8	38.2	53.5	67.6	76.7	90.5	100.0

Figure 5.5: Calculations for No.1 mixture

For No. 2 mixture with the max size of coarse aggregate 19 mm, we have the calculations according to figure 5.6.

			Max size of agg (mm)	Value for n						Max Size of agg (mm)	Value for n
			19	0.25						19	0.35

0.15 mm	0.3 mm	0.6 mm	1.18 mm	2.36 mm	4.75 mm	9.5 mm	12.5 mm	19 mm	25 mm	Sieve size
4.6	10.5	18.0	27.3	39.5	55.1	74.8	84.1	100.0	111.8	n1 value
6.3	13.9	22.8	33.2	45.8	60.9	78.8	86.7	100.0	109.5	n2 value

Percent (%)	Type	Aggregate
23	Crushed	Coarse Gravel
11	Crushed	Fina Gravel
33	Natural	Sand No.1
33	Crushed	Sand No.2
100	◦◦◦◦◦	Total

0.15 mm	0.3 mm	0.6 mm	1.18 mm	2.36 mm	4.75 mm	9.5 mm	12.5 mm	19 mm	25 mm	Aggregate
0	0	0	0	0	0.6	2.2	11.0	22.8	23.0	Coarse Gravel
0	0	0	0	0.0	0.5	6.4	10.4	11.0	11	Fine Gravel
2.0	8.0	12.7	16.6	22.0	27.8	33	33	33	33	Sand No.1
2.0	4.6	8.1	12.8	21.5	32.2	33	33	33	33	Sand No.2
4.0	12.6	20.8	29.4	43.4	61.2	74.6	87.4	99.8	100.0	Total

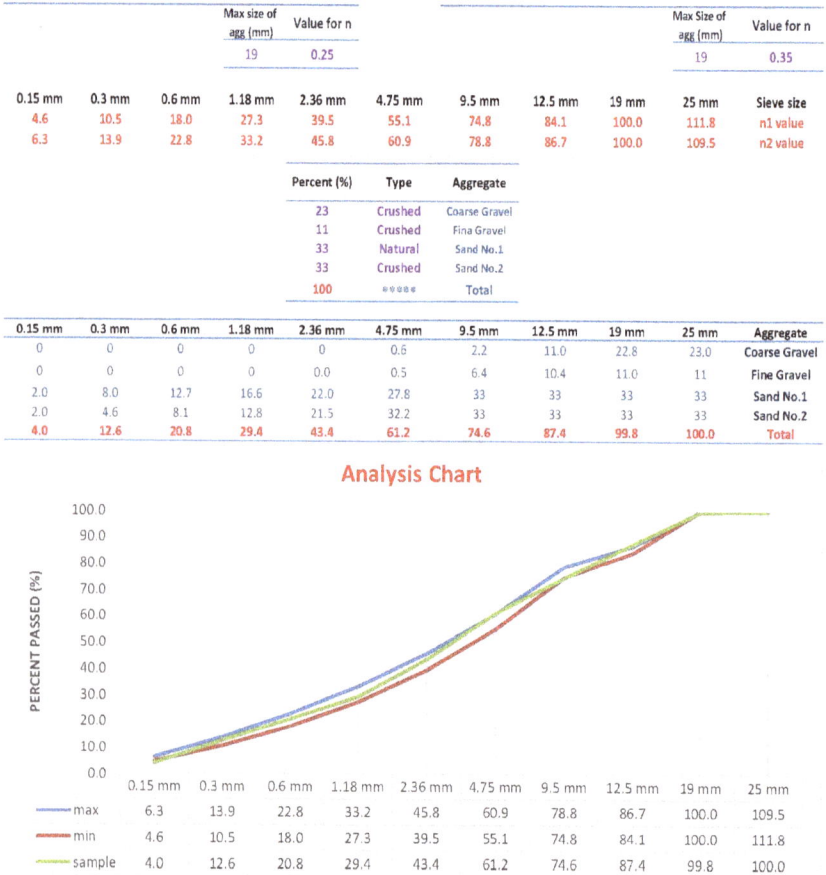

Analysis Chart

	0.15 mm	0.3 mm	0.6 mm	1.18 mm	2.36 mm	4.75 mm	9.5 mm	12.5 mm	19 mm	25 mm
max	6.3	13.9	22.8	33.2	45.8	60.9	78.8	86.7	100.0	109.5
min	4.6	10.5	18.0	27.3	39.5	55.1	74.8	84.1	100.0	111.8
sample	4.0	12.6	20.8	29.4	43.4	61.2	74.6	87.4	99.8	100.0

Figure 5.6: Calculations for No.2 mixture

For No.3 mixture with the max size of coarse aggregate 12.5 mm, we have the calculations according to figure 5.7.

									Max size of agg (mm)	Value for n
									12.5	0.25

									Max Size of agg (mm)	Value for n
									12.5	0.35

0.15 mm	0.3 mm	0.6 mm	1.18 mm	2.36 mm	4.75 mm	9.5 mm	12.5 mm	19 mm	25 mm	Sieve size
5.5	12.5	21.4	32.5	46.9	65.5	89.0	100.0	118.9	133.0	n1 value
7.3	16.0	26.3	38.2	52.8	70.2	90.8	100.0	115.3	126.2	n2 value

Percent (%)	Type	Aggregate
0	Crushed	Coarse Gravel
26	Crushed	Fina Gravel
37	Natural	Sand No.1
37	Crushed	Sand No.2
100	≈≈≈≈≈	Total

0.15 mm	0.3 mm	0.6 mm	1.18 mm	2.36 mm	4.75 mm	9.5 mm	12.5 mm	19 mm	25 mm	Aggregate
0	0	0	0	0	0.0	0.0	0.0	0.0	0.0	Coarse Gravel
0	0	0	0	0.0	1.3	15.1	24.7	26.0	26	Fine Gravel
2.2	8.9	14.3	18.6	24.6	31.2	37	37	37	37	Sand No.1
2.2	5.2	9.0	14.4	24.1	36.1	37	37	37	37	Sand No.2
4.4	14.1	23.3	32.9	48.7	68.6	89.1	98.7	100.0	100.0	Total

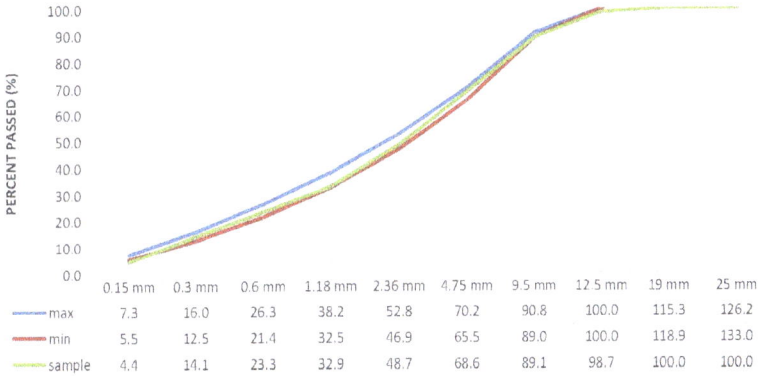

Analysis Chart

	0.15 mm	0.3 mm	0.6 mm	1.18 mm	2.36 mm	4.75 mm	9.5 mm	12.5 mm	19 mm	25 mm
max	7.3	16.0	26.3	38.2	52.8	70.2	90.8	100.0	115.3	126.2
min	5.5	12.5	21.4	32.5	46.9	65.5	89.0	100.0	118.9	133.0
sample	4.4	14.1	23.3	32.9	48.7	68.6	89.1	98.7	100.0	100.0

Figure 5.7: Calculations for No.3 mixture.

As you can see in the pictures above, we have three tables for three concrete mixtures as tables 5.6 to 5.8.

Table 5.6: Percentage of each aggregate for No.1 mixture

Aggregate	Type	Percent
Coarse gravel	12-25 Crushed	27%
Fine gravel	5-12 Crushed	15%
Sand	0-8 Natural	29%
Sand	0-5 Crushed	29%

Table 5.7: Percentage of each aggregate for No. 2 mixture

Aggregate	Type	Percent
Coarse gravel	11-19 Crushed	23%
Fine gravel	5-12 Crushed	11%
Sand	0-8 Natural	33%
Sand	0-5 Crushed	33%

Table 5.8: Percentage of each aggregate for No. 3 mixture

Aggregate	Type	Percent
Fine gravel	5-12 Crushed	26%
Sand	0-8 Natural	37%
Sand	0-5 Crushed	37%

The reason for the equality of two kinds of sand is the possibility of the batching plants of this ready mixed plant for mixing different sands.

5.3.4 Step (4): Specify fineness module of total aggregates

The definition of total aggregates fineness module and the excel file give us the following figures for each mixture.

Total	0.15 mm	0.3 mm	0.6 mm	1.18 mm	2.36 mm	4.75 mm	9.5 mm	19 mm	37.5 mm	Sieve Size
•••••	7.6	7.2	7.5	12.3	15.3	14.1	22.9	9.5	0.0	Percent remained
491.6	96.5	88.9	81.7	74.2	61.8	46.5	32.4	9.5	0.0	Cumulative percent remained
					4.92	Total Fineness Module				

Figure 5.8: Calculation of fineness module for mixture No.1

Total	0.15 mm	0.3 mm	0.6 mm	1.18 mm	2.36 mm	4.75 mm	9.5 mm	19 mm	37.5 mm	Sieve Size
•••••	8.6	8.2	8.6	14.1	17.7	13.4	25.3	0.2	0.0	Percent remained
454.3	96.0	87.4	79.2	70.6	56.6	38.8	25.4	0.2	0.0	Cumulative percent remained
					4.54	Total Fineness Module				

Figure 5.9: Calculation of fineness module for mixture No.2

Total	0.15 mm	0.3 mm	0.6 mm	1.18 mm	2.36 mm	4.75 mm	9.5 mm	19 mm	37.5 mm	Sieve Size
*****	9.7	9.2	9.6	15.8	19.9	20.5	10.9	0.0	0.0	Percent remained
418.8	95.6	85.9	76.7	67.1	51.3	31.4	10.9	0.0	0.0	Cumulative percent remained
					4.19	Total Fineness Module				

Figure 5.10: Calculation of fineness module for mixture No.3

For fineness module of each concrete mixture, we have table 5.9.

Table 5.9: Fineness module for each mixture	
Mixture number	**Fineness module**
No.1	4.92
No.2	4.54
No.3	4.19

As mentioned before, mixture No.1 is the coarser mixture for the plant and mixture No.3 is the finer mixture which we are going to use.

5.3.5 Step (5): Specify water to binder ratio

As mentioned before, this step is one of the most important steps in the concrete mix design. Making the right decision in this step will guaranty our mix design to be accurate.

We have to consider that we have two kinds of cements in our plant. As mentioned before, in this ready mixed factory, there are two batching plants. In one batching, we are going to use company (B) cement with the standard compressive strength of 550 kg/cm^2, which shows that this cement is an I-525 cement. In the other one, we are going to use company (A) cement with the standard compressive strength of 470 kg/cm^2, which shows that this cement is stronger than I-425 and weaker than I-525 cement.

To use tables 4.4 and 4.5 we must consider the percentage of crushed and natural aggregates in the total mixture. So, we will have table 5.10 for that.

Table 5.10: Percentage of crushed and natural aggregates		
Mixture	**Percentage of crushed aggregate**	**Percentage of natural aggregate**
Mixture No.1	71%	29%
Mixture No.2	67%	33%
Mixture No.3	63%	37%

According to table 5.10 and for simplicity, we use 70% of crushed and 30% of natural aggregates.

To estimate w/b ratio for two kinds of cements, we have tables 5.11 and 5.12.

Table 5.11: Estimation of w/b ratio for I-525 cement

Type of concrete	f_{cm}	w/b Table 4.4	w/b Table 4.6	w/b Table 4.7	w/b mean value
C25	30	0.64	0.61	-----	0.62
C30	35	0.59	0.53	-----	0.56
C35	40	0.54	0.47	-----	0.50
C40	45	0.51	0.43	0.42	0.45
C45	50	0.47	-----	0.40	0.43
C50	55	0.42	-----	0.39	0.40
C60	65	-----	-----	0.36	0.36
C70	75	-----	-----	0.33	0.33

Table 5.12: Estimation of w/b ratio for I-425 cement

Type of concrete	f_{cm}	w/b Table 4.4	w/b Table 4.6	w/b Table 4.7	w/b mean value
C25	30	0.58	0.61	-----	0.59
C30	35	0.53	0.53	-----	0.53
C35	40	0.49	0.47	-----	0.48
C40	45	0.45	0.43	0.38	0.42
C45	50	0.39	-----	0.36	0.38
C50	55	0.35	-----	0.35	0.35
C60	65	-----	-----	0.32	0.32
C70	75	-----	-----	0.29	0.29

For company (A) and (B) cements, according to their compressive strength, we must use linear correlation to estimate w/b ratio. So, we will have table 5.13 for the final w/b ratio for two cements of the plant.

Table 5.13: Final amounts of w/b for company (A) and (B) cements

Type of concrete	w/b for company (B) cement	w/b for company (A) cement
C25	0.63	0.60
C30	0.57	0.54
C35	0.50	0.49
C40	0.46	0.43
C45	0.44	0.40
C50	0.41	0.37
C60	0.37	0.34
C70	0.34	0.31

5.3.6   Step (6): Specify free water in concrete

For this step, we should first define a target slump for each kind of concrete. In ready mixed plants, we often use the same production slump for all kinds of concrete. Therefore, we can again use one target slump for all concrete types.

As defined before, to estimate the "n" value for total aggregates, we used 140 to 150 mm slump. But since it takes about 45 to 90 minutes to transport our concrete from the plant to different projects, we should consider higher slump for our production plant.

According to table 3.1, we will define target slump as 190 mm (table 5.14). After the transportation of concrete and reduction of slump, we will have a 140 to 170 mm slump, depending on the transportation time of the project.

Table 5.14: Target production slump for the ready mixed plant	
Target slump for all concrete types	**190 mm**

Now consider tables 4.9 and 4.10 to estimate the amount of free water in the concretes. To use these tables, we assume the amount of cement (or total binder) for different concretes. To do that, we have table 5.15.

Table 5.15: Assumed amount of cement for different kinds of concrete		
Type of concrete	**Minimum amount of binder (kg)**	**Maximum amount of binder (kg)**
C25	300	325
C30	325	400
C35	350	425
C40	375	425
C45	400	450
C50	425	475
C60	450	500
C70	475	525

According to table 5.15 for the defined amount of binder, table 5.9 for the fineness module of each mix and the target slump, we use table 5.16 for the free water in each concrete for target slump of 90 mm and table 5.17 for target slump of 150 mm and table 5.18 for final free water with target slump of 190 mm.

Table 5.16: Free water for each type of concrete for target slump of 90 mm

Type	Min Binder (kg)	Max Binder (kg)	F module of aggregates	Min Free Water (L)	Max Free Water (L)	Mean Free Water (L)
C25.1	300	350	4.9	183	191	**187**
C25.2	300	350	4.5	196	204	**200**
C30.1	325	400	4.5	200	212	**206**
C30.2	325	400	4.2	210	222	**216**
C30.3	325	400	4.5	200	212	**206**
C30.4	325	400	4.2	210	222	**216**
C35.1	350	425	4.5	204	216	**210**
C35.2	350	425	4.2	214	226	**220**
C35.3	350	425	4.5	204	216	**210**
C35.4	350	425	4.2	214	226	**220**
C40.1	375	450	4.5	208	220	**214**
C40.2	375	450	4.2	218	230	**224**
C40.3	375	450	4.5	208	220	**214**
C40.4	375	450	4.2	218	230	**224**
C45.1	400	475	4.2	222	234	**228**
C45.2	400	475	4.2	222	234	**228**
C50.1	425	500	4.2	226	238	**232**
C50.2	425	500	4.2	226	238	**232**
C60	450	525	4.2	230	242	**236**
C70	475	550	4.2	234	246	**240**

Table 5.17: Free water for each type of concrete for target slump of 150 mm

Type	Min Binder (kg)	Max Binder (kg)	F module of aggregates	Min Free Water (L)	Max Free Water (L)	Mean Free Water (L)
C25.1	300	350	4.9	199	207	**203**
C25.2	300	350	4.5	213	221	**217**
C30.1	325	400	4.5	217	229	**223**
C30.2	325	400	4.2	228	240	**234**
C30.3	325	400	4.5	217	229	**223**
C30.4	325	400	4.2	228	240	**234**
C35.1	350	425	4.5	221	233	**227**
C35.2	350	425	4.2	232	244	**238**
C35.3	350	425	4.5	221	233	**227**
C35.4	350	425	4.2	232	244	**238**
C40.1	375	450	4.5	225	237	**231**
C40.2	375	450	4.2	236	248	**242**
C40.3	375	450	4.5	225	237	**231**
C40.4	375	450	4.2	236	248	**242**
C45.1	400	475	4.2	240	252	**246**
C45.2	400	475	4.2	240	252	**246**
C50.1	425	500	4.2	244	256	**250**
C50.2	425	500	4.2	244	256	**250**
C60	450	425	4.2	248	260	**254**
C70	475	550	4.2	252	264	**258**

Table 5.18: Final free water for each type of concrete and target slump of 190 mm

Concrete Type	Free water for target slump:90 mm	Free water for target slump:150 mm	Final free water for target slump:190 mm
C25.1	187	203	214
C25.2	200	217	228
C30.1	206	223	234
C30.2	216	234	246
C30.3	206	223	234
C30.4	216	234	246
C35.1	210	227	238
C35.2	220	238	250
C35.3	210	227	238
C35.4	220	238	250
C40.1	214	231	242
C40.2	224	242	254
C40.3	214	231	242
C40.4	224	242	254
C45.1	228	246	258
C45.2	228	246	258
C50.1	232	250	262
C50.2	232	250	262
C60	236	254	266
C70	240	258	270

So, as you can see, table 5.18 is the final amounts of free water for all kinds of concrete in this plant.

5.3.6.1 Using of super-plasticizer

In this part, we must consider different types of cement that we are going to use because the effect of the super-plasticizer on different kinds of cement is different according to part 2.5.4.

Another fact that we have to account for is the amount of super-plasticizer we are going to use for each concrete according to Appendix 3.
As we have the same target slump for all concrete, and we are using different kinds of binders for different kinds of concrete, we can use more super-plasticizer for higher strengths.

Also, we cannot have too many different variants to decide on the dosage of super-plasticizer. So, we will use the same dosage of super-plasticizer for two types of Portland cement. According to their specification, we know that concretes made with company (B) cement need more water for the same slump.

Table 5.19 shows suggested dosage of our super-plasticizer.

Table 5.19: Suggested dosage of BPC-40 for different kinds of concrete

Type of concrete	Suggested dosage	water reduction rate for company (A) cement	Water reduction rate for company (B) cement
C25.1	0.3%	15%	14%
C25.2	0.3%	15%	14%
C30.1	0.5%	19%	18%
C30.2	0.5%	19%	18%
C30.3	0.5%	19%	18%
C30.4	0.5%	19%	18%
C35.1	0.6%	21%	20%
C35.2	0.6%	21%	20%
C35.3	0.6%	21%	20%
C35.4	0.6%	21%	20%
C40.1	0.7%	23%	21%
C40.2	0.7%	23%	21%
C40.3	0.7%	23%	21%
C40.4	0.7%	23%	21%
C45.1	0.8%	25%	23%
C45.2	0.8%	25%	23%
C50.1	0.9%	27%	25%
C50.2	0.9%	27%	25%
C60	1.1%	30%	28%
C70	1.3%	33%	31%

3.5.7 *Step (7): Specify the amount of cement and other binders*

In this step, we work on each type of concrete separately. For each one, we have the water to binder ratio from table 5.13, amount of free water from table 5.18 and the dosage of super plasticizer and water reduction rate for each cement, from table 5.19. So, we can calculate the amount of total binder. For some kinds of concretes the total binder will actually be pure Portland cement. But for the others, in which we are going to use some kind of pozzolans, we first specify the amount of GGBS and/or silica fume from table 2.12 and then we increase the amount of total binder according to table 4.11. Finally, we calculate the amount of Portland cement, GGBS and/or silica fume.

- C25.1 company (A) cement: w/b=0.6   free water=214L

When we use 0.3% of super-plasticizer, the water reduction rate is 15% and reduced water is 182 L:
b=182/0.6 = 303 kg   SP= 303 x 0.3% = 0.91 kg
In this case the total binder is pure Portland cement.

- C25.1 company (B) cement   w/b=0.63   free water=214 L

When we use 0.3% of super-plasticizer, the water reduction rate is 14% and reduced water is 184 L:
b=184/0.63 = 292 kg   SP= 292 x 0.3% = 0.88 kg
In this case total binder is pure Portland cement.

- C25.2 company (A) cement: w/b=0.6   free water=228 L

When we use 0.3% of super-plasticizer, the water reduction rate is 15% and reduced water is 194 L:
b=194/0.6 = 323kg   SP= 323 x 0.3% = 0.97kg
In this case total binder is pure Portland cement.

- C25.2 Company(B) cement: w/b=0.63   free water=228L

When we use 0.3% of super-plasticizer, the water reduction rate is 14% and reduced water is 196 L:
b=196/0.63 = 311 kg   SP= 311 x 0.3% = 0.93 kg
In this case, the total binder is pure Portland cement.

- C30.1 company (A) cement: w/b=0.54   free water=234 L

When we use 0.5% of super-plasticizer, the water reduction rate is 19% and reduced water is 189 L:
b=189/0.54 = 350 kg   SP= 350 x 0.5% = 1.75 kg
In this case, the total binder is pure Portland cement.

- C30.1 company (B) cement: w/b=0.57   free water=234 L

When we use 0.5% of super-plasticizer, the water reduction rate is 18% and reduced water is 192 L:
b=192/0.57 = 337 kg   SP= 337 x 0.5% = 1.69 kg
In this case, the total binder is pure Portland cement.

- C30.2 company (A) cement: w/b=0.54   free water=246 L

When we use 0.5% of super-plasticizer, the water reduction rate is 19% and reduced water is 199 L:
b=199/0.54 = 368 kg   SP= 368 x 0.5% = 1.84 kg
In this case, the total binder is pure Portland cement.

- C30.2 company (B) cement: w/b=0.57   free water=246 L

When we use 0.5% of super-plasticizer, the water reduction rate is 18% and reduced water is 202 L:
b=202/0.57 = 355 kg   SP= 355 x 0.5% = 1.78 kg

In this case, the total binder is pure Portland cement.

- C30.3 company (A) cement + GGBS   w/b=0.54   free water=234 L

When we use 0.5% of super-plasticizer, the water reduction rate is 19% and reduced water is 189 L:
b=189/0.54 = 350 kg
According to table 2.12, we decided to use 20% of GGBS. You can see from table 4.11 that we should increase the total binder by 15% for this case.
Increased b=350 x 1.15 = 402 kg
GGBS= 402 x 20% = 80 kg   C= 402-80= 322 kg   SP=402 x 0.5%= 2.01 kg

- C30.3 company (B) cement + GGBS   w/b=0.57   free water=234 L

When we use 0.5% of super-plasticizer, the water reduction rate is 18% and reduced water is 192 L:
b=192/0.57 = 337 kg
According to table 2.12, we decided to use 20% of GGBS. You can see from table 4.11 that we should increase the total binder by 15% for this case.
Increased b=337 x 1.15 = 388 kg
GGBS= 388 x 20% = 78 kg   C= 388-78= 310 kg   SP=388 x 0.5%= 1.94 kg

- C30.4 company (A) cement + GGBS   w/b=0.54   free water=246 L

When we use 0.5% of super-plasticizer, the water reduction rate is 19% and reduced water is 199 L:
b=199/0.54 = 368 kg
According to table 2.12, we decided to use 20% of GGBS. You can see from table 4.11 that we should increase the total binder by 15% for this case.
Increased b=368 x 1.15 = 423 kg
GGBS= 423 x 20% = 85 kg   C= 423-85= 338 kg   SP=423 x 0.5%= 2.11 kg

- C30.4 company (B) cement + GGBS   w/b=0.57   free water=246 L

When we use 0.5% of super-plasticizer, the water reduction rate is 18% and reduced water is 202 L:
b=202/0.57 = 355 kg
According to table 2.12, we decided to use 20% of GGBS. You can see from table 4.11 that we should increase the total binder by 15% for this case.
Increased b=355 x 1.15 = 408 kg
GGBS= 408 x 20% = 82 kg   C= 408-82= 326 kg   SP=408 x 0.5%= 2.04 kg

- C35.1 company (A) cement: w/b=0.49   free water= 238 L

When we use 0.6% of super-plasticizer, the water reduction rate is 21% and reduced water will be 188 L
b=188/0.49 = 384 kg   SP= 384 x 0.6% = 2.30 kg
In this case the total binder is pure Portland cement.

- C35.1 Company(B) cement: w/b=0.50   free water=238 L

When we use 0.6% of super-plasticizer, the water reduction rate is 20% and reduced water is 190 L:
b=190/0.50 = 380 kg   SP= 380 x 0.6% = 2.28 kg
In this case, the total binder is pure Portland cement.

- C35.2 company (A) cement: w/b=0.49   free water=250 L

When we use 0.6% of super-plasticizer, the water reduction rate is 21% and reduced water is 197 L:
b=197/0.49 = 402 kg   SP= 402 x 0.6% = 2.41 kg
In this case, the total binder is pure Portland cement.

- C35.2 company (B) cement: w/b=0.50   free water=250 L

When we use 0.6% of super-plasticizer, the water reduction rate is 20% and reduced water is 200 L:
b=200/0.50 = 400kg   SP= 400 x 0.6% = 2.40kg
In this case, the total binder is pure Portland cement.

- C35.3 company (A) cement + GGBS: w/b=0.49   free water=238 L

When we use 0.6% of super-plasticizer, the water reduction rate is 21% and reduced water is 188 L:
b=188/0.49 = 384 kg
According to table 2.12, we decided to use 20% of GGBS. You can see from table 4.11 that we should increase the total binder by 15% for this case.
Increased b=384 x 1.15 = 442 kg
GGBS= 442 x 20% = 88 kg   C= 442-88= 354 kg   SP=442 x 0.6%= 2.65 kg

- C35.3 company (B) cement + GGBS: w/b=0.50   free water=238 L

When we use 0.6% of super-plasticizer, the water reduction rate is 20% and reduced water is 190 L:
b=190/0.50 = 380 kg
According to table 2.12, we decided to use 20% of GGBS. You can see from table 4.11 that we should increase the total binder by 15% for this case.
Increased b=380 x 1.15 = 437 kg

GGBS= 437 x 20% = 87 kg   C= 437-87= 350 kg   SP=437 x 0.6%= 2.62 kg

- C35.4 company (A) cement + GGBS: w/b=0.49   free water=250 L

When we use 0.6% of super-plasticizer, the water reduction rate is 21% and reduced water is 197 L:

b=197/0.49 = 402 kg

According to table 2.12, we decided to use 20% of GGBS. You can see from table 4.11 that we should increase the total binder by 15% for this case.

Increased b=402 x 1.15 = 462 kg

GGBS= 462 x 20% = 92 kg   C= 462-92= 370 kg   SP=462 x 0.6%= 2.77 kg

- C35.4 company (B) cement + GGBS: w/b=0.50   free water=250 L

When we use 0.6% of super-plasticizer, the water reduction rate is 20% and reduced water is 200 L:

b=200/0.50 = 400 kg

According to table 2.12, we decided to use 20% of GGBS. You can see from table 4.11 that we should increase the total binder by 15% for this case.

Increased b=400 x 1.15 = 460 kg

GGBS= 460 x 20% = 92 kg   C= 460-92= 368 kg   SP=460 x 0.6%= 2.76 kg

- C40.1 company (A) cement: w/b=0.43   free water=242 L

When we use 0.7% of super-plasticizer, the water reduction rate is 23% and reduced water is 186 L.

b=186/0.43 = 433 kg   SP= 433 x 0.7% = 3.03 kg

In this case, the total binder is pure Portland cement.

- C40.1 company (B) cement: w/b=0.46   free water=242 L

When we use 0.7% of super-plasticizer, the water reduction rate is 21% and reduced water is 191 L:

b=191/0.46 = 415 kg   SP= 415 x 0.7% = 2.91 kg

In this case, the total binder is pure Portland cement.

- C40.2 Company(A) cement: w/b=0.43   free water=254 L

When we use 0.7% of super-plasticizer the water reduction rate is 23% and reduced water is 196 L:

b=196/0.43 = 456 kg   SP= 456 x 0.7% = 3.19 kg

In this case, the total binder is pure Portland cement.

- C40.2 company (B) cement: w/b=0.46   free water=254 L

When we use 0.7% of super-plasticizer, the water reduction rate will be 21% and reduced water will be 201 L.
b=201/0.46 = 437 kg   SP= 437 x 0.7% = 3.06 kg
In this case, the total binder is pure Portland cement.

- C40.3 company (A) cement + GGBS: w/b=0.43   free water=242 L

When we use 0.7% of super-plasticizer, the water reduction rate is 23% and reduced water is 186 L.
b=186/0.43 = 433 kg
According to table 2.12, we decided to use 20% of GGBS. You can see from table 4.11 that we should increase the total binder by15% for this case.
Increased b=433 x 1.15 = 498 kg
GGBS= 498 x 20% = 100 kg   C= 498-100= 398 kg   SP=498 x 0.7%= 3.49 kg

- C40.3 company (B) cement + GGBS: w/b=0.46   free water=242 L

When we use 0.7% of super-plasticizer, the water reduction rate is 21% and reduced water is 191 L:
b=191/0.46 = 415 kg
According to table 2.12, we decided to use 20% of GGBS. You can see from table 4.11 that we should increase the total binder by 15% for this case.
Increased b=415 x 1.15 = 477 kg
GGBS= 477 x 20% = 95 kg   C= 477-95= 382 kg   SP=477 x 0.7%= 3.34 kg

- C40.4 company (A) cement + GGBS: w/b=0.43   free water=254 L

When we use 0.7% of super-plasticizer, the water reduction rate is 23% and reduced water is 196 L:
b=196/0.43 = 456 kg
According to table 2.12, we decided to use 20% of GGBS. You canl see from table 4.11 that we should increase the total binder by 15% for this case.
Increased b=456 x 1.15 = 524 kg
GGBS= 524 x 20% = 105 kg   C= 524-105= 419 kg   SP=524 x 0.7%= 3.67 kg

- C40.4 company (B) cement + GGBS: w/b=0.46   free water=254 L

When we use 0.7% of super-plasticizer, the water reduction rate is 21% and reduced water is 201 L:
b=201/0.46 = 437 kg
According to table 2.12, we decided to use 20% of GGBS. You can see from table 4.11 that we should increase the total binder by 15% for this case.
Increased b=437 x 1.15 = 502 kg

GGBS= 502 x 20% = 100 kg   C= 502-100= 402 kg   SP=502 x 0.7%= 3.51 kg

- C45.1 company (A) cement: w/b=0.40   free water=258 L

When we use 0.8% of super-plasticizer, the water reduction rate is 25% and reduced water is 193 L:
b=193/0.40 = 482 kg   SP= 482 x 0.8% = 3.86 kg
In this case, the total binder is pure Portland cement.

- C45.1 company (B) cement: w/b=0.44   free water=258 L

When we use 0.8% of super-plasticizer, the water reduction rate is 23% and reduced water is 199 L:
b=199/0.44 = 453 kg   SP= 453 x 0.8% = 3.62 kg
In this case, the total binder is pure Portland cement.

- C45.2 company (A) cement + GGBS: w/b=0.40   free water=258 L

When we use 0.8% of super-plasticizer, the water reduction rate is 25% and reduced water is 193 L:
b=193/0.40 = 482 kg
According to table 2.12, we decided to use 20% of GGBS. You can see from table 4.11 that we should increase the total binder by 15% for this case.
Increased b=482 x 1.15 = 554 kg
GGBS= 554 x 20% = 111 kg   C= 554-111= 443 kg   SP=554 x 0.8%= 4.43 kg

- C45.2 company (B) cement + GGBS: w/b=0.44   free water=258 L

When we use 0.8% of super-plasticizer, the water reduction rate will be 23% and reduced water will be 199 L.
b=199/0.44 = 453 kg
According to table 2.12, we decided to use 20% of GGBS. You can see from table 4.11 that we should increase the total binder by 15% for this case.
Increased b=453 x 1.15 = 521 kg
GGBS= 521 x 20% = 104 kg   C= 521-104= 417 kg   SP=521 x 0.8%= 4.17 kg

- C50.1 company (A) cement + GGBS: w/b=0.37   free water=262 L

When we use 0.9% of super-plasticizer, the water reduction rate is 27% and reduced water is 191 L:
b=191/0.37 = 516 kg
According to table 2.12, we decided to use 20% of GGBS. You can see from table 4.11 that we should increase the total binder by 15% for this case.
Increased b=516 x 1.15 = 593 kg
GGBS= 593 x 20% = 119 kg   C= 593-119= 474 kg   SP=593 x 0.9%= 5.34 kg

- C50.1 company (B) cement + GGBS: w/b=0.41 free water=262 L

When we use 0.9% of super-plasticizer the water reduction rate is 25% and reduced water is 196 L:

b=196/0.41 = 478 kg

According to table, 2.12 we decided to use 20% of GGBS. You can see from table 4.11 that we should increase the total binder by 15% for this case.

Increased b=478 x 1.15 = 550 kg

GGBS= 550 x 20% = 110 kg   C= 550-110= 440 kg   SP=550 x 0.9%= 4.95 kg

- C50.2 company (A) cement + GGBS+SF: w/b=0.37 free water=262 L

When we use 0.9% of super-plasticizer, the water reduction rate is 27% and reduced water is 191 L:

b=191/0.37 = 516 kg

According to table 2.12, we decided to use 20% of GGBS and 5% of silica fume. You can see from table 4.11 that we should increase the total binder by 15% for this case. So:

Increased b=516 x 1.15 = 593 kg

GGBS= 593 x 20% = 119 kg   SF=593 x 5% = 30 kg   C= 593-119- 30 = 444 kg

SP=593 x 0.9%= 5.34 kg

- C50.2 company (B) cement + GGBS + SF: w/b=0.41 free water=262 L

When we use 0.9% of super-plasticizer, the water reduction rate is 25% and reduced water is 196 L:

b=196/0.41 = 478 kg

According to table 2.12, we decided to use 20% of GGBS and 5% of silica fume. You can see from table 4.11 that we should increase the total binder by 15% for this case.

Increased b=478 x 1.15 = 550 kg

GGBS= 550 x 20% = 110 kg   SF=550 x 5% = 27 kg   C= 550-110- 27 = 413 kg

SP=550 x 0.9%= 4.95 kg

- C60 company (A) cement + GGBS + SF: w/b=0.34 free water=266 L

When we use 1.1% of super-plasticizer, the water reduction rate is 30% and reduced water is 186 L:

b=186/0.34 = 547 kg

According to table 2.12, we decided to use 20% of GGBS and 7% of silica fume. You can see from table 4.11 that we should increase the total binder by 15% for this case.

Increased b=547 x 1.15 = 629 kg

GGBS= 629 x 20% = 126 kg   SF=629 x 7% = 44 kg   C= 629-126- 44 = 459 kg

SP=629 x 1.1%= 6.92 kg

- C60 company (B) cement + GGBS + SF: w/b=0.37   free water=266 L

When we use 1.1% of super-plasticizer, the water reduction rate is 28% and reduced water is 191 L:

b=191/0.37 = 516 kg

According to table 2.12, we decided to use 20% of GGBS and 7% of silica fume. You can see from table 4.11 that we should increase the total binder by 15% for this case.

Increased b=516 x 1.15 = 593 kg

GGBS= 593 x 20% = 119 kg   SF=593 x 7% = 42 kg   C= 593-119- 42 = 432 kg

SP=593 x 1.1%= 6.52 kg

- C70 company (A) cement + GGBS + SF: w/b=0.31   free water=270 L

When we use 1.3% of super-plasticizer, the water reduction rate is 33% and reduced water is 181 L:

b=181/0.31 = 584 kg

According to table 2.12, we decided to use 20% of GGBS and 9% of silica fume. You can see from table 4.11 that we should increase the total binder by15% for this case.

Increased b=584 x 1.15 = 672 kg

GGBS= 672 x 20% = 134 kg   SF=672 x 9% = 61 kg   C= 672-134- 61 = 477 kg

SP=672 x 1.3%= 8.74 kg

- C70 company (B) cement + GGBS + SF: w/b=0.34   free water=270 L

When we use 1.3% of super-plasticizer, the water reduction rate is 31% and reduced water is 186 L.

b=186/0.34 = 547 kg

According to table 2.12, we decided to use 20% of GGBS and 9% of silica fume. You can see from table 4.11 that we should increase the total binder by 15% for this case.

Increased b=547 x 1.15 = 629 kg

GGBS= 629 x 20% = 126 kg   SF=629 x 9% = 57 kg   C= 629-126- 57 = 446 kg

SP=629 x 1.3%= 8.18 kg

After finalizing our calculations, you can see the final amounts of superplasticizer, final free water for SSD aggregates, Portland cement, GGBS and silica fume for concretes made with company (A) cement in table 5.20 and concretes made with company (B) cement in table 5.21.

Table 5.20: Final amounts of binders, super-plasticizer and water for concretes made with company (A) cement

Concrete Type	Super-plasticizer Dosage/Amount	Water for SSD aggregates	Amount of Portland cement	Amount of GGBS	Amount of silica fume
C25.1	0.3%=0.91 kg	182 L	303 kg	0	0
C25.2	0.3%=0.97 kg	194 L	323 kg	0	0
C30.1	0.5%=1.75 kg	189 L	350 kg	0	0
C30.2	0.5%=1.84 kg	199 L	368 kg	0	0
C30.3	0.5%=2.01 kg	189 L	322 kg	80 kg	0
C30.4	0.5%=2.11 kg	199 L	338 kg	85 kg	0
C35.1	0.6%=2.30 kg	188 L	384 kg	0	0
C35.2	0.6%=2.41 kg	197 L	402 kg	0	0
C35.3	0.6%=2.65 kg	188 L	354 kg	88 kg	0
C35.4	0.6%=2.77 kg	197 L	370 kg	92 kg	0
C40.1	0.7%=3.03 kg	186 L	433 kg	0	0
C40.2	0.7%=3.19 kg	196 L	465 kg	0	0
C40.3	0.7%=3.49 kg	186 L	398 kg	100 kg	0
C40.4	0.7%=3.67 kg	196 L	419 kg	105 kg	0
C45.1	0.8%=3.86 kg	193 L	482 kg	0	0
C45.2	0.8%=4.43 kg	193 L	443 kg	111 kg	0
C50.1	0.9%=5.34 kg	191 L	474 kg	119 kg	0
C50.2	0.9%=5.34 kg	191 L	444 kg	119 kg	30 kg
C60	1.1%=6.92 kg	186 L	459 kg	126 kg	44 kg
C70	1.3%=8.74 kg	181 L	477 kg	134 kg	61 kg

Table 5.21: Final amounts of binders, super-plasticizer and water for concretes made with company (B) cement

Concrete Type	Super-plasticizer Dosage/Amount	Water for SSD aggregates	Amount of Portland cement	Amount of GGBS	Amount of silica fume
C25.1	0.3%=0.88 kg	184 L	292 kg	0	0
C25.2	0.3%=0.93 kg	196 L	311 kg	0	0
C30.1	0.5%=1.69 kg	192 L	337 kg	0	0
C30.2	0.5%=1.78 kg	202 L	355 kg	0	0
C30.3	0.5%=1.94 kg	192 L	310 kg	78 kg	0
C30.4	0.5%=2.04 kg	202 L	326 kg	82 kg	0
C35.1	0.6%=2.28 kg	190 L	380 kg	0	0
C35.2	0.6%=2.40 kg	200 L	400 kg	0	0
C35.3	0.6%=2.62 kg	190 L	350 kg	87 kg	0
C35.4	0.6%=2.76 kg	200 L	368 kg	92 kg	0
C40.1	0.7%=2.91 kg	191 L	415 kg	0	0
C40.2	0.7%=3.06 kg	201 L	437 kg	0	0
C40.3	0.7%=3.34 kg	191 L	382 kg	95 kg	0
C40.4	0.7%=3.51 kg	201 L	402 kg	100 kg	0
C45.1	0.8%=3.62 kg	199 L	453 kg	0	0
C45.2	0.8%=4.17 kg	199 L	417 kg	104 kg	0

C50.1	0.9%=4.95 kg	196 L	440 kg	110 kg	0
C50.2	0.9%=4.95 kg	196 L	413 kg	110 kg	27 kg
C60	1.1%=6.52 kg	191 L	432 kg	119 kg	42 kg
C70	1.3%=8.18 kg	186 L	446 kg	126 kg	57 kg

5.3.8   Step (8): Specify the total volume of aggregates in concrete

For this step, we use equation 4.7 and for this equation, we have all the data from tables 5.20 and 5.21 instead of the air volume in concrete.

To estimate the amount of air in the concrete, we use table 4.12 for the maximum size of coarse aggregate and table 4.13 for the chemical base of super-plasticizer.

We have many kinds of concretes with different maximum size of coarse aggregates, but all kinds of concrete are made with a polycarboxylate ether type super-plasticizer with different dosages. You should know that higher dosage of super-plasticizer will cause more air in concrete. But we can simplify our calculations in this step.

So according to the tables 4.12 and 4.13, we will assume 1.2% of air in concretes made with the maximum size of coarse aggregate 25 mm, and 1.5% of air in concrete made with maximum size of coarse aggregate 19 mm and 1.8% of air in concrete made with maximum size of coarse aggregate 12 mm. You can see it in the table 5.22.

Table 5.22: Percentage and amount of air in concretes of the ready mixed plant

Max size of coarse aggregate	Types of concrete	Percentage of entrapped air	Amount of air in $1m^3$
25mm	C25.1	1.2%	12L
19mm	C25.2 C30.1, C30.3 C35.1, C35.3 C40.1, C40.3	1.5%	15L
12.5mm	C30.2, C30.4 C35.2, C35.4 C40.2, C40.4 C45.1, C45.2 C50.1, C50.2 C60 C70	1.8%	18L

According to these descriptions, we have the calculations for the total volume of aggregates in concretes made with company (A) cement as:

- C25.1: V=1000 - (303/3.16) - (182) - (12) - (0.91/1.08) = 709 L
- C25.2: V=1000 - (323/3.16) - (194) - (15) - (0.97/1.08) = 688 L
- C30.1: V=1000 - (350/3.16) - (189) - (15) - (1.75/1.08) = 684 L

- C30.2: $V=1000 - (368/3.16) - (199) - (18) - (1.84/1.08) = 665$ L
- C30.3: $V=1000 - (322/3.16) - (80/2.9) - (189) - (15) - (2.01/1.08) = 665$ L
- C30.4: $V=1000 - (338/3.16) - (85/2.9) - (199) - (18) - (2.11/1.08) = 645$ L
- C35.1: $V=1000 - (384/3.16) - (188) - (15) - (2.3/1.08) = 673$ L
- C35.2: $V=1000 - (402/3.16) - (197) - (18) - (2.41/1.08) = 656$ L
- C35.3: $V=1000 - (354/3.16) - (88/2.9) - (188) - (15) - (2.65/1.08) = 653$ L
- C35.4: $V=1000 - (370/3.16) - (92/2.9) - (197) - (18) - (2.77/1.08) = 634$ L
- C40.1: $V=1000 - (433/3.16) - (186) - (15) - (3.03/1.08) = 660$ L
- C40.2: $V=1000 - (465/3.16) - (196) - (18) - (3.19/1.08) = 636$ L
- C40.3: $V=1000 - (398/3.16) - (100/2.9) - (186) - (15) - (3.49/1.08) = 635$ L
- C40.4: $V=1000 - (419/3.16) - (105/2.9) - (196) - (18) - (3.67/1.08) = 614$ L
- C45.1: $V=1000 - (482/3.16) - (193) - (18) - (3.86/1.08) = 633$ L
- C45.2: $V=1000 - (443/3.16) - (111/2.9) - (193) - (18) - (4.43/1.08) = 606$ L
- C50.1: $V=1000 - (474/3.16) - (119/2.9) - (191) - (18) - (5.34/1.08) = 595$ L
- C50.2: $V=1000-(444/3.16) - (119/2.9) - (30/2.25) -(191)-(18)-(5.34/1.08)=592$ L
- C60: $V=1000-(459/3.16) - (126/2.9) - (44/2.25) - (186) - (18) - (6.92/1.08) =582$ L
- C70: $V=1000-(477/3.16) -(134/2.9) -(61/2.25) -(181) -(18) -(8.74/1.08) =569$ L

And we have the calculations for the total volume of aggregates in concretes made with company (B) cement as:

- C25.1: $V=1000 - (292/3.15) - (184) - (12) - (0.88/1.08) = 711$ L
- C25.2: $V=1000 - (311/3.15) - (196) - (15) - (0.93/1.08) = 690$ L
- C30.1: $V=1000 - (337/3.15) - (192) - (15) - (1.69/1.08) = 684$ L
- C30.2: $V=1000 - (355/3.15) - (202) - (18) - (1.78/1.08) = 665$ L
- C30.3: $V=1000 - (310/3.15) - (78/2.9) - (192) - (15) - (1.94/1.08) = 666$ L
- C30.4: $V=1000 - (326/3.15) - (82/2.9) - (202) - (18) - (2.04/1.08) = 646$ L
- C35.1: $V=1000 - (380/3.15) - (190) - (15) - (2.28/1.08) = 672$ L
- C35.2: $V=1000 - (400/3.15) - (200) - (18) - (2.28/1.08) = 653$ L
- C35.3: $V=1000 - (350/3.15) - (87/2.9) - (190) - (15) - (2.62/1.08) = 652$ L
- C35.4: $V=1000 - (368/3.15) - (92/2.9) - (200) - (18) - (2.76/1.08) = 631$ L
- C40.1: $V=1000 - (415/3.15) - (191) - (15) - (2.91/1.08) = 660$ L
- C40.2: $V=1000 - (437/3.15) - (201) - (18) - (3.06/1.08) = 640$ L
- C40.3: $V=1000 - (382/3.15) - (95/2.9) - (191) - (15) - (3.34/1.08) = 637$ L
- C40.4: $V=1000 - (402/3.15) - (100/2.9) - (201) - (18) - (3.51/1.08) = 616$ L
- C45.1: $V=1000 - (453/3.15) - (199) - (18) - (3.62/1.08) = 636$ L
- C45.2: $V=1000 - (417/3.15) - (104/2.9) - (199) - (18) - (4.17/1.08) = 611$ L
- C50.1: $V=1000 - (440/3.15) - (110/2.9) - (196) - (18) - (4.95/1.08) = 604$ L
- C50.2: $V=1000-(413/3.15) -(110/2.9) -(27/2.25) - (196) -(18) -(4.95/1.08) =601$ L

- C60: V=1000-(432/3.15) - (119/2.9) -(42/2.25) -(191) -(18) -(6.52/1.08) =589 L
- C70: V=1000-(446/3.15) -(126/2.9) -(57/2.25) -(186) -(18) -(8.18/1.08) =578 L

Finally, you can see total volume of aggregates for all kinds of concretes in table 5.23.

Table 5.23: Total volume of aggregates in different concretes

Concrete Type	Total volume of aggregates for concretes made with company (A) cement (L)	Total volume of aggregates for concretes made with company (B) cement (L)
C25.1	709	711
C25.2	688	690
C30.1	684	684
C30.2	665	665
C30.3	665	666
C30.4	645	646
C35.1	673	672
C35.2	656	653
C35.3	653	652
C35.4	634	631
C40.1	660	660
C40.2	636	640
C40.3	635	637
C40.4	614	616
C45.1	633	636
C45.2	606	611
C50.1	595	604
C50.2	592	601
C60	582	589
C70	569	578

5.3.9   Step (9): Calculating the weight of aggregates in saturated surface dry (SSD) condition

For this step you can use tables 5.24 to 5.63.

Table 5.24: Aggregates weight in $1m^3$ for C25.1 company (A) cement

Aggregates type	Percentage (step three)	Aggregates volume (L)	SSD Specific gravity (kg/L)	Weight of aggregate (kg)
Gravel 12-25	27%	191	2.795	535
Gravel 5-12	15%	106	2.785	296
Sand 0-8	29%	206	2.699	555
Sand 0-5	29%	206	2.714	558
Total	100%	709	-----	1944

Table 5.25: Aggregates weight in 1m3 for C25.1 company (B) cement

Aggregates type	Percentage (step three)	Aggregates volume (L)	SSD Specific gravity (kg/L)	Weight of aggregate (kg)
Gravel 12-25	27%	192	2.795	537
Gravel 5-12	15%	107	2.785	297
Sand 0-8	29%	206	2.699	557
Sand 0-5	29%	206	2.714	560
Total	100%	711	-----	1950

Table 5.26: Aggregates weight in 1m3 for C25.2 company (A) cement

Aggregates type	Percentage (step three)	Aggregates volume (L)	SSD Specific gravity (kg/L)	Weight of aggregate (kg)
Gravel 11-19	23%	158	2.791	442
Gravel 5-12	11%	76	2.785	211
Sand 0-8	33%	227	2.699	613
Sand 0-5	33%	227	2.714	616
Total	100%	688	-----	1881

Table 5.27: Aggregates weight in 1m3 for C25.2 company (B) cement

Aggregates type	Percentage (step three)	Aggregates volume (L)	SSD Specific gravity (kg/L)	Weight of aggregate (kg)
Gravel 11-19	23%	159	2.791	443
Gravel 5-12	11%	76	2.785	211
Sand 0-8	33%	228	2.699	615
Sand 0-5	33%	228	2.714	618
Total	100%	690	-----	1887

Table 5.28: Aggregates weight in 1m3 for C30.1 company (A) cement

Aggregates type	Percentage (step three)	Aggregates volume (L)	SSD Specific gravity (kg/L)	Weight of aggregate (kg)
Gravel 11-19	23%	157	2.791	439
Gravel 5-12	11%	75	2.785	210
Sand 0-8	33%	226	2.699	609
Sand 0-5	33%	226	2.714	613
Total	100%	684	-----	1870

Table 5.29: Aggregates weight in 1m3 for C30.1 company (B) cement

Aggregates type	Percentage (step three)	Aggregates volume (L)	SSD Specific gravity (kg/L)	Weight of aggregate (kg)
Gravel 11-19	23%	157	2.791	439
Gravel 5-12	11%	75	2.785	210
Sand 0-8	33%	226	2.699	609
Sand 0-5	33%	226	2.714	613
Total	100%	684	-----	1870

Table 5.30: Aggregates weight in 1m3 for C30.2 company (A) cement

Aggregates type	Percentage (step three)	Aggregates volume (L)	SSD Specific gravity (kg/L)	Weight of aggregate (kg)
Gravel 5-12	26%	173	2.785	482
Sand 0-8	37%	246	2.699	664
Sand 0-5	37%	246	2.714	668
Total	100%	665	-----	1813

Table 5.31: Aggregates weight in 1m3 for C30.2 company (B) cement

Aggregates type	Percentage (step three)	Aggregates volume (L)	SSD Specific gravity (kg/L)	Weight of aggregate (kg)
Gravel 5-12	26%	173	2.785	482
Sand 0-8	37%	246	2.699	664
Sand 0-5	37%	246	2.714	668
Total	100%	665	-----	1813

Table 5.32: Aggregates weight in 1m3 for C30.3 company (A) cement

Aggregates type	Percentage (step three)	Aggregates volume (L)	SSD Specific gravity (kg/L)	Weight of aggregate (kg)
Gravel 11-19	23%	153	2.791	427
Gravel 5-12	11%	73	2.785	204
Sand 0-8	33%	219	2.699	592
Sand 0-5	33%	219	2.714	596
Total	100%	665	-----	1818

Table 5.33: Aggregates weight in 1m3 for C30.3 company (B) cement

Aggregates type	Percentage (step three)	Aggregates volume (L)	SSD Specific gravity (kg/L)	Weight of aggregate (kg)
Gravel 11-19	23%	153	2.791	428
Gravel 5-12	11%	73	2.785	204
Sand 0-8	33%	220	2.699	593
Sand 0-5	33%	220	2.714	596
Total	100%	666	-----	1821

Table 5.34: Aggregates weight in 1m3 for C30.4 company (A) cement

Aggregates type	Percentage (step three)	Aggregates volume (L)	SSD Specific gravity (kg/L)	Weight of aggregate (kg)
Gravel 5-12	26%	168	2.785	467
Sand 0-8	37%	239	2.699	644
Sand 0-5	37%	239	2.714	648
Total	100%	645	-----	1759

Table 5.35: Aggregates weight in 1m3 for C30.4 company (B) cement

Aggregates type	Percentage (step three)	Aggregates volume (L)	SSD Specific gravity (kg/L)	Weight of aggregate (kg)
Gravel 5-12	26%	168	2.785	468
Sand 0-8	37%	239	2.699	645
Sand 0-5	37%	239	2.714	649
Total	100%	646	-----	1762

Table 5.36: Aggregates weight in 1m3 for C35.1 company (A) cement

Aggregates type	Percentage (step three)	Aggregates volume (L)	SSD Specific gravity (kg/L)	Weight of aggregate (kg)
Gravel 11-19	23%	155	2.791	432
Gravel 5-12	11%	74	2.785	206
Sand 0-8	33%	222	2.699	599
Sand 0-5	33%	222	2.714	603
Total	100%	673	-----	1840

Table 5.37: Aggregates weight in 1m3 for C35.1 company (B) cement

Aggregates type	Percentage (step three)	Aggregates volume (L)	SSD Specific gravity (kg/L)	Weight of aggregate (kg)
Gravel 11-19	23%	155	2.791	431
Gravel 5-12	11%	74	2.785	206
Sand 0-8	33%	222	2.699	599
Sand 0-5	33%	222	2.714	602
Total	100%	672	-----	1838

Table 5.38: Aggregates weight in 1m3 for C35.2 company (A) cement

Aggregates type	Percentage (step three)	Aggregates volume (L)	SSD Specific gravity (kg/L)	Weight of aggregate (kg)
Gravel 5-12	26%	171	2.785	475
Sand 0-8	37%	243	2.699	655
Sand 0-5	37%	243	2.714	659
Total	100%	656	-----	1789

Table 5.39: Aggregates weight in 1m3 for C35.2 company (B) cement

Aggregates type	Percentage (step three)	Aggregates volume (L)	SSD Specific gravity (kg/L)	Weight of aggregate (kg)
Gravel 5-12	26%	170	2.785	473
Sand 0-8	37%	242	2.699	652
Sand 0-5	37%	242	2.714	656
Total	100%	653	-----	1781

Table 5.40: Aggregates weight in 1m3 for C35.3 company (A) cement

Aggregates type	Percentage (step three)	Aggregates volume (L)	SSD Specific gravity (kg/L)	Weight of aggregate (kg)
Gravel 11-19	23%	150	2.791	419
Gravel 5-12	11%	72	2.785	200
Sand 0-8	33%	215	2.699	582
Sand 0-5	33%	215	2.714	585
Total	100%	653	-----	1786

Table 5.41: Aggregates weight in 1m3 for C35.3 company (B) cement

Aggregates type	Percentage (step three)	Aggregates volume (L)	SSD Specific gravity (kg/L)	Weight of aggregate (kg)
Gravel 11-19	23%	150	2.791	419
Gravel 5-12	11%	72	2.785	200
Sand 0-8	33%	215	2.699	581
Sand 0-5	33%	215	2.714	584
Total	100%	652	-----	1783

Table 5.42: Aggregates weight in 1m3 for C35.4 company (A) cement

Aggregates type	Percentage (step three)	Aggregates volume (L)	SSD Specific gravity (kg/L)	Weight of aggregate (kg)
Gravel 5-12	26%	165	2.785	459
Sand 0-8	37%	235	2.699	633
Sand 0-5	37%	235	2.714	637
Total	100%	634	-----	1729

Table 5.43: Aggregates weight in 1m3 for C35.4 company (B) cement

Aggregates type	Percentage (step three)	Aggregates volume (L)	SSD Specific gravity (kg/L)	Weight of aggregate (kg)
Gravel 5-12	26%	164	2.785	457
Sand 0-8	37%	233	2.699	630
Sand 0-5	37%	233	2.714	634
Total	100%	631	-----	1721

Table 5.44: Aggregates weight in 1m3 for C40.1 company (A) cement

Aggregates type	Percentage (step three)	Aggregates volume (L)	SSD Specific gravity (kg/L)	Weight of aggregate (kg)
Gravel 11-19	23%	152	2.791	424
Gravel 5-12	11%	73	2.785	202
Sand 0-8	33%	218	2.699	588
Sand 0-5	33%	218	2.714	591
Total	100%	660	-----	1805

Materials Research Forum LLC
https://doi.org/10.21741/9781644900598

Table 5.45: Aggregates weight in 1m3 for C40.1 company (B) cement

Aggregates type	Percentage (step three)	Aggregates volume (L)	SSD Specific gravity (kg/L)	Weight of aggregate (kg)
Gravel 11-19	23%	152	2.791	424
Gravel 5-12	11%	73	2.785	202
Sand 0-8	33%	218	2.699	588
Sand 0-5	33%	218	2.714	591
Total	100%	660	-----	1805

Table 5.46: Aggregates weight in 1m3 for C40.2 company (A) cement

Aggregates type	Percentage (step three)	Aggregates volume (L)	SSD Specific gravity (kg/L)	Weight of aggregate (kg)
Gravel 5-12	26%	165	2.785	461
Sand 0-8	37%	235	2.699	635
Sand 0-5	37%	235	2.714	639
Total	100%	636	-----	1734

Table 5.47: Aggregates weight in 1m3 for C40.2 company (B) cement

Aggregates type	Percentage (step three)	Aggregates volume (L)	SSD Specific gravity (kg/L)	Weight of aggregate (kg)
Gravel 5-12	26%	166	2.785	463
Sand 0-8	37%	237	2.699	639
Sand 0-5	37%	237	2.714	643
Total	100%	640	-----	1745

Table 5.48: Aggregates weight in 1m3 for C40.3 company (A) cement

Aggregates type	Percentage (step three)	Aggregates volume (L)	SSD Specific gravity (kg/L)	Weight of aggregate (kg)
Gravel 11-19	23%	146	2.794	408
Gravel 5-12	11%	70	2.785	195
Sand 0-8	33%	210	2.699	566
Sand 0-5	33%	210	2.714	569
Total	100%	635	-----	1736

Table 5.49: Aggregates weight in 1m3 for C40.3 company (B) cement

Aggregates type	Percentage (step three)	Aggregates volume (L)	SSD Specific gravity (kg/L)	Weight of aggregate (kg)
Gravel 11-19	23%	147	2.791	409
Gravel 5-12	11%	70	2.785	195
Sand 0-8	33%	210	2.699	567
Sand 0-5	33%	210	2.714	571
Total	100%	637	-----	1742

Table 5.50: Aggregates weight in 1m3 for C40.4 company (A) cement

Aggregates type	Percentage (step three)	Aggregates volume (L)	SSD Specific gravity (kg/L)	Weight of aggregate (kg)
Gravel 5-12	26%	160	2.785	445
Sand 0-8	37%	227	2.699	613
Sand 0-5	37%	227	2.714	617
Total	100%	614	-----	1674

Table 5.51: Aggregates weight in 1m3 for C40.4 company (B) cement

Aggregates type	Percentage (step three)	Aggregates volume (L)	SSD Specific gravity (kg/L)	Weight of aggregate (kg)
Gravel 5-12	26%	160	2.785	446
Sand 0-8	37%	228	2.699	615
Sand 0-5	37%	228	2.714	619
Total	100%	616	-----	1680

Table 5.52: Aggregates weight in 1m3 for C45.1 company (A) cement

Aggregates type	Percentage (step three)	Aggregates volume (L)	SSD Specific gravity (kg/L)	Weight of aggregate (kg)
Gravel 5-12	26%	165	2.785	458
Sand 0-8	37%	234	2.699	632
Sand 0-5	37%	234	2.714	636
Total	100%	633	-----	1726

Table 5.53: Aggregates weight in 1m3 for C45.1 company (B) cement

Aggregates type	Percentage (step three)	Aggregates volume (L)	SSD Specific gravity (kg/L)	Weight of aggregate (kg)
Gravel 5-12	26%	165	2.785	461
Sand 0-8	37%	235	2.699	635
Sand 0-5	37%	235	2.714	639
Total	100%	636	-----	1734

Table 5.54: Aggregates weight in 1m3 for C45.2 company (A) cement

Aggregates type	Percentage (step three)	Aggregates volume (L)	SSD Specific gravity (kg/L)	Weight of aggregate (kg)
Gravel 5-12	26%	158	2.785	439
Sand 0-8	37%	224	2.699	605
Sand 0-5	37%	224	2.714	609
Total	100%	606	-----	1653

Table 5.55: Aggregates weight in 1m3 for C45.2 company (B) cement

Aggregates type	Percentage (step three)	Aggregates volume (L)	SSD Specific gravity (kg/L)	Weight of aggregate (kg)
Gravel 5-12	26%	159	2.785	442
Sand 0-8	37%	226	2.699	610
Sand 0-5	37%	226	2.714	614
Total	100%	611	-----	1666

Table 5.56: Aggregates weight in 1m3 for C50.1 company (A) cement

Aggregates type	Percentage (step three)	Aggregates volume (L)	SSD Specific gravity (kg/L)	Weight of aggregate (kg)
Gravel 5-12	26%	155	2.785	431
Sand 0-8	37%	220	2.699	594
Sand 0-5	37%	220	2.714	597
Total	100%	595	-----	1623

Table 5.57: Aggregates weight in 1m3 for C50.1 company (B) cement

Aggregates type	Percentage (step three)	Aggregates volume (L)	SSD Specific gravity (kg/L)	Weight of aggregate (kg)
Gravel 5-12	26%	157	2.785	437
Sand 0-8	37%	223	2.699	603
Sand 0-5	37%	223	2.714	607
Total	100%	604	-----	1647

Table 5.58: Aggregates weight in 1m3 for C50.2 company (A) cement

Aggregates type	Percentage (step three)	Aggregates volume (L)	SSD Specific gravity (kg/L)	Weight of aggregate (kg)
Gravel 5-12	26%	154	2.785	429
Sand 0-8	37%	219	2.699	591
Sand 0-5	37%	219	2.714	594
Total	100%	592	-----	1614

Table 5.59: Aggregates weight in 1m3 for C50.2 company (B) cement

Aggregates type	Percentage (step three)	Aggregates volume (L)	SSD Specific gravity (kg/L)	Weight of aggregate (kg)
Gravel 5-12	26%	156	2.785	435
Sand 0-8	37%	222	2.699	600
Sand 0-5	37%	222	2.714	604
Total	100%	601	-----	1639

Table 5.60: Aggregates weight in 1m3 for C60 company (A) cement

Aggregates type	Percentage (step three)	Aggregates volume (L)	SSD Specific gravity (kg/L)	Weight of aggregate (kg)
Gravel 5-12	26%	151	2.785	421
Sand 0-8	37%	215	2.699	581
Sand 0-5	37%	215	2.714	584
Total	100%	582	-----	1587

Table 5.61: Aggregates weight in 1m3 for C60 company (B) cement

Aggregates type	Percentage (step three)	Aggregates volume (L)	SSD Specific gravity (kg/L)	Weight of aggregate (kg)
Gravel 5-12	26%	153	2.785	426
Sand 0-8	37%	218	2.699	588
Sand 0-5	37%	218	2.714	591
Total	100%	589	-----	1606

Table 5.62: Aggregates weight in 1m3 for C70 company (A) cement

Aggregates type	Percentage (step three)	Aggregates volume (L)	SSD Specific gravity (kg/L)	Weight of aggregate (kg)
Gravel 5-12	26%	148	2.785	412
Sand 0-8	37%	211	2.699	568
Sand 0-5	37%	211	2.714	571
Total	100%	569	-----	1552

Table 5.63: Aggregates weight in 1m3 for C70 company (B) cement

Aggregates type	Percentage (step three)	Aggregates volume (L)	SSD Specific gravity (kg/L)	Weight of aggregate (kg)
Gravel 5-12	26%	150	2.785	418
Sand 0-8	37%	214	2.699	577
Sand 0-5	37%	214	2.714	581
Total	100%	578	-----	1576

5.3.10 Step (10): Calculating the real weight of aggregates and water in the concrete

For accurately checking our mix designs, we will dry all of the aggregates in an oven and then use them for checking the mix design in the laboratory. For this reason, we have tables 5.64 to 5.103.

Table 5.64: Weight of dry aggregates for C25.1 company (A) cement

Aggregates Type	SSD weight (kg)	Water absorption (%)	Amount of water (kg)	Weight of dry aggregate (kg)
Gravel 12-25	535	0.96	5.1	530
Gravel 5-12	296	0.98	2.9	293
Sand 0-8	555	1.69	9.4	546
Sand 0-5	558	1.54	8.6	549
Total	1944	-----	26.0	1918

Table 5.65: Weight of dry aggregates for C25.1 company (B) cement

Aggregates Type	SSD weight (kg)	Water absorption (%)	Amount of water (kg)	Weight of dry aggregate (kg)
Gravel 12-25	537	0.96	5.2	532
Gravel 5-12	297	0.98	2.9	294
Sand 0-8	557	1.69	9.4	548
Sand 0-5	560	1.54	8.6	551
Total	1950	-----	26.1	1925

Table 5.66: Weight of dry aggregates for C25.2 company (A) cement

Aggregates Type	SSD weight (kg)	Water absorption (%)	Amount of water (kg)	Weight of dry aggregate (kg)
Gravel 11-19	442	0.98	4.3	438
Gravel 5-12	211	0.98	2.1	209
Sand 0-8	613	1.69	10.4	603
Sand 0-5	616	1.54	9.5	607
Total	1881	-----	26.2	1857

Table 5.67: Weight of dry aggregates for C25.2 company (B) cement

Aggregates Type	SSD weight (kg)	Water absorption (%)	Amount of water (kg)	Weight of dry aggregate (kg)
Gravel 11-19	443	0.98	4.3	439
Gravel 5-12	211	0.98	2.1	209
Sand 0-8	615	1.69	10.4	605
Sand 0-5	618	1.54	9.5	608
Total	1887	-----	26.3	1861

Table 5.68: Weight of dry aggregates for C30.1 company (A) cement

Aggregates Type	SSD weight (kg)	Water absorption (%)	Amount of water (kg)	Weight of dry aggregate (kg)
Gravel 11-19	439	0.98	4.3	435
Gravel 5-12	210	0.98	2.1	208
Sand 0-8	609	1.69	10.3	599
Sand 0-5	613	1.54	9.4	604
Total	1870	-----	26.1	1846

Table 5.69: Weight of dry aggregates for C30.1 company (B) cement

Aggregates Type	SSD weight (kg)	Water absorption (%)	Amount of water (kg)	Weight of dry aggregate (kg)
Gravel 11-19	439	0.98	4.3	435
Gravel 5-12	210	0.98	2.1	208
Sand 0-8	609	1.69	10.3	599
Sand 0-5	613	1.54	9.4	604
Total	1870	-----	26.1	1846

Table 5.70: Weight of dry aggregates for C30.2 company (A) cement

Aggregates Type	SSD weight (kg)	Water absorption (%)	Amount of water (kg)	Weight of dry aggregate (kg)
Gravel 5-12	482	0.98	4.7	477
Sand 0-8	664	1.69	11.2	653
Sand 0-5	668	1.54	10.3	658
Total	1813	-----	26.2	1788

Table 5.71: Weight of dry aggregates for C30.2 company (B) cement

Aggregates Type	SSD weight (kg)	Water absorption (%)	Amount of water (kg)	Weight of dry aggregate (kg)
Gravel 5-12	482	0.98	4.7	477
Sand 0-8	664	1.69	11.2	653
Sand 0-5	668	1.54	10.3	658
Total	1813	-----	26.2	1788

Table 5.72: Weight of dry aggregates for C30.3 company (A) cement

Aggregates Type	SSD weight (kg)	Water absorption (%)	Amount of water (kg)	Weight of dry aggregate (kg)
Gravel 11-19	427	0.98	4.2	423
Gravel 5-12	204	0.98	2.0	202
Sand 0-8	592	1.69	10.0	582
Sand 0-5	596	1.54	9.2	587
Total	1818	-----	25.4	1794

Table 5.73: Weight of dry aggregates for C30.3 company (B) cement

Aggregates Type	SSD weight (kg)	Water absorption (%)	Amount of water (kg)	Weight of dry aggregate (kg)
Gravel 11-19	428	0.98	4.2	424
Gravel 5-12	204	0.98	2.0	202
Sand 0-8	593	1.69	10.0	583
Sand 0-5	596	1.54	9.2	587
Total	1821	-----	25.4	1796

Table 5.74: Weight of dry aggregates for C30.4 company (A) cement

Aggregates Type	SSD weight (kg)	Water absorption (%)	Amount of water (kg)	Weight of dry aggregate (kg)
Gravel 5-12	467	0.98	4.6	462
Sand 0-8	644	1.69	10.9	633
Sand 0-5	648	1.54	10.0	638
Total	1759	-----	25.4	1734

Table 5.75: Weight of dry aggregates for C30.4 company (B) cement

Aggregates Type	SSD weight (kg)	Water absorption (%)	Amount of water (kg)	Weight of dry aggregate (kg)
Gravel 5-12	468	0.98	4.6	463
Sand 0-8	645	1.69	10.9	634
Sand 0-5	649	1.54	10.0	639
Total	1762	-----	25.5	1737

Materials Research Forum LLC
https://doi.org/10.21741/9781644900598

Table 5.76: Weight of dry aggregates for C35.1 company (A) cement

Aggregates Type	SSD weight (kg)	Water absorption (%)	Amount of water (kg)	Weight of dry aggregate (kg)
Gravel 11-19	432	0.98	4.2	428
Gravel 5-12	206	0.98	2.0	204
Sand 0-8	599	1.69	10.1	589
Sand 0-5	603	1.54	9.3	594
Total	1840	-----	25.7	1814

Table 5.77: Weight of dry aggregates for C35.1 company (B) cement

Aggregates Type	SSD weight (kg)	Water absorption (%)	Amount of water (kg)	Weight of dry aggregate (kg)
Gravel 11-19	431	0.98	4.2	427
Gravel 5-12	206	0.98	2.0	204
Sand 0-8	599	1.69	10.1	589
Sand 0-5	602	1.54	9.3	593
Total	1838	-----	25.6	1812

Table 5.78: Weight of dry aggregates for C35.2 company (A) cement

Aggregates Type	SSD weight (kg)	Water absorption (%)	Amount of water (kg)	Weight of dry aggregate (kg)
Gravel 5-12	475	0.98	4.7	470
Sand 0-8	655	1.69	11.1	644
Sand 0-5	659	1.54	10.1	649
Total	1789	-----	25.9	1763

Table 5.79: Weight of dry aggregates for C35.2 company (B) cement

Aggregates Type	SSD weight (kg)	Water absorption (%)	Amount of water (kg)	Weight of dry aggregate (kg)
Gravel 5-12	473	0.98	4.6	468
Sand 0-8	652	1.69	11.0	641
Sand 0-5	656	1.54	10.1	646
Total	1781	-----	25.8	1755

Table 5.80: Weight of dry aggregates for C35.3 company (A) cement

Aggregates Type	SSD weight (kg)	Water absorption (%)	Amount of water (kg)	Weight of dry aggregate (kg)
Gravel 11-19	419	0.98	4.1	415
Gravel 5-12	200	0.98	2.0	198
Sand 0-8	582	1.69	9.8	572
Sand 0-5	585	1.54	9.0	576
Total	1786	-----	24.9	1761

Table 5.81: Weight of dry aggregates for C35.3 company (B) cement

Aggregates Type	SSD weight (kg)	Water absorption (%)	Amount of water (kg)	Weight of dry aggregate (kg)
Gravel 11-19	419	0.98	4.1	415
Gravel 5-12	200	0.98	2.0	198
Sand 0-8	581	1.69	9.8	571
Sand 0-5	584	1.54	9.0	575
Total	1783	-----	24.9	1759

Table 5.82: Weight of dry aggregates for C35.4 company (A) cement

Aggregates Type	SSD weight (kg)	Water absorption (%)	Amount of water (kg)	Weight of dry aggregate (kg)
Gravel 5-12	459	0.98	4.5	455
Sand 0-8	633	1.69	10.7	622
Sand 0-5	637	1.54	9.8	627
Total	1729	-----	25.0	1704

Table 5.83: Weight of dry aggregates for C35.4 company (B) cement

Aggregates Type	SSD weight (kg)	Water absorption (%)	Amount of water (kg)	Weight of dry aggregate (kg)
Gravel 5-12	457	0.98	4.5	453
Sand 0-8	630	1.69	10.6	619
Sand 0-5	634	1.54	9.8	624
Total	1721	-----	24.9	1696

Table 5.84: Weight of dry aggregates for C40.1 company (A) cement

Aggregates Type	SSD weight (kg)	Water absorption (%)	Amount of water (kg)	Weight of dry aggregate (kg)
Gravel 11-19	424	0.98	4.2	420
Gravel 5-12	202	0.98	2.0	200
Sand 0-8	588	1.69	9.9	578
Sand 0-5	591	1.54	9.1	582
Total	1805	-----	25.2	1780

Table 5.85: Weight of dry aggregates for C40.1 company (B) cement

Aggregates Type	SSD weight (kg)	Water absorption (%)	Amount of water (kg)	Weight of dry aggregate (kg)
Gravel 11-19	424	0.98	4.2	420
Gravel 5-12	202	0.98	2.0	200
Sand 0-8	588	1.69	9.9	578
Sand 0-5	591	1.54	9.1	582
Total	1805	-----	25.2	1780

Table 5.86: Weight of dry aggregates for C40.2 company (A) cement				
Aggregates Type	**SSD weight (kg)**	**Water absorption (%)**	**Amount of water (kg)**	**Weight of dry aggregate (kg)**
Gravel 5-12	461	0.98	4.5	456
Sand 0-8	635	1.69	10.7	624
Sand 0-5	639	1.54	9.8	629
Total	1734	-----	25.1	1710

Table 5.87: Weight of dry aggregates for C40.2 company (B) cement				
Aggregates Type	**SSD weight (kg)**	**Water absorption (%)**	**Amount of water (kg)**	**Weight of dry aggregate (kg)**
Gravel 5-12	463	0.98	4.5	458
Sand 0-8	639	1.69	10.8	628
Sand 0-5	643	1.54	9.9	633
Total	1745	-----	25.2	1720

Table 5.88: Weight of dry aggregates for C40.3 company (A) cement				
Aggregates Type	**SSD weight (kg)**	**Water absorption (%)**	**Amount of water (kg)**	**Weight of dry aggregate (kg)**
Gravel 11-19	408	0.98	4.0	404
Gravel 5-12	195	0.98	1.9	193
Sand 0-8	566	1.69	9.6	556
Sand 0-5	569	1.54	8.8	560
Total	1736	-----	24.2	1714

Table 5.89: Weight of dry aggregates for C40.3 company (B) cement				
Aggregates Type	**SSD weight (kg)**	**Water absorption (%)**	**Amount of water (kg)**	**Weight of dry aggregate (kg)**
Gravel 11-19	409	0.98	4.0	405
Gravel 5-12	195	0.98	1.9	193
Sand 0-8	567	1.69	9.6	557
Sand 0-5	571	1.54	8.8	562
Total	1742	-----	24.3	1718

Table 5.90: Weight of dry aggregates for C40.4 company (A) cement				
Aggregates Type	**SSD weight (kg)**	**Water absorption (%)**	**Amount of water (kg)**	**Weight of dry aggregate (kg)**
Gravel 5-12	445	0.98	4.4	441
Sand 0-8	613	1.69	10.4	603
Sand 0-5	617	1.54	9.5	607
Total	1674	-----	24.2	1651

Table 5.91: Weight of dry aggregates for C40.4 company (B) cement

Aggregates Type	SSD weight (kg)	Water absorption (%)	Amount of water (kg)	Weight of dry aggregate (kg)
Gravel 5-12	446	0.98	4.4	442
Sand 0-8	615	1.69	10.4	605
Sand 0-5	619	1.54	9.5	609
Total	1680	------	24.3	1656

Table 5.92: Weight of dry aggregates for C45.1 company (A) cement

Aggregates Type	SSD weight (kg)	Water absorption (%)	Amount of water (kg)	Weight of dry aggregate (kg)
Gravel 5-12	458	0.98	4.5	454
Sand 0-8	632	1.69	10.7	621
Sand 0-5	636	1.54	9.8	626
Total	1726	------	25.0	1701

Table 5.93: Weight of dry aggregates for C45.1 company (B) cement

Aggregates Type	SSD weight (kg)	Water absorption (%)	Amount of water (kg)	Weight of dry aggregate (kg)
Gravel 5-12	461	0.98	4.5	456
Sand 0-8	635	1.69	10.7	624
Sand 0-5	639	1.54	9.8	629
Total	1734	------	25.1	1710

Table 5.94: Weight of dry aggregates for C45.2 company (A) cement

Aggregates Type	SSD weight (kg)	Water absorption (%)	Amount of water (kg)	Weight of dry aggregate (kg)
Gravel 5-12	439	0.98	4.3	435
Sand 0-8	605	1.69	10.2	595
Sand 0-5	609	1.54	9.4	600
Total	1653	------	23.9	1629

Table 5.95: Weight of dry aggregates for C45.2 company (B) cement

Aggregates Type	SSD weight (kg)	Water absorption (%)	Amount of water (kg)	Weight of dry aggregate (kg)
Gravel 5-12	442	0.98	4.3	438
Sand 0-8	610	1.69	10.3	600
Sand 0-5	614	1.54	9.5	605
Total	1666	------	24.1	1642

Table 5.96: Weight of dry aggregates for C50.1 company (A) cement

Aggregates Type	SSD weight (kg)	Water absorption (%)	Amount of water (kg)	Weight of dry aggregate (kg)
Gravel 5-12	431	0.98	4.2	427
Sand 0-8	594	1.69	10.0	584
Sand 0-5	597	1.54	9.2	588
Total	1623	------	23.5	1599

Table 5.97: Weight of dry aggregates for C50.1 company (B) cement

Aggregates Type	SSD weight (kg)	Water absorption (%)	Amount of water (kg)	Weight of dry aggregate (kg)
Gravel 5-12	437	0.98	4.3	433
Sand 0-8	603	1.69	10.2	593
Sand 0-5	607	1.54	9.3	598
Total	1647	-----	23.8	1623

Table 5.98: Weight of dry aggregates for C50.2 company (A) cement

Aggregates Type	SSD weight (kg)	Water absorption (%)	Amount of water (kg)	Weight of dry aggregate (kg)
Gravel 5-12	429	0.98	4.2	425
Sand 0-8	591	1.69	10.0	581
Sand 0-5	594	1.54	9.1	585
Total	1614	-----	23.3	1591

Table 5.99: Weight of dry aggregates for C50.2 company (B) cement

Aggregates Type	SSD weight (kg)	Water absorption (%)	Amount of water (kg)	Weight of dry aggregate (kg)
Gravel 5-12	435	0.98	4.3	431
Sand 0-8	600	1.69	10.1	590
Sand 0-5	604	1.54	9.3	595
Total	1639	-----	23.7	1615

Table 5.100: Weight of dry aggregates for C60 company (A) cement

Aggregates Type	SSD weight (kg)	Water absorption (%)	Amount of water (kg)	Weight of dry aggregate (kg)
Gravel 5-12	421	0.98	4.1	417
Sand 0-8	581	1.69	9.8	571
Sand 0-5	584	1.54	9.0	575
Total	1587	-----	22.9	1563

Table 5.101: Weight of dry aggregates for C60 company (B) cement

Aggregates Type	SSD weight (kg)	Water absorption (%)	Amount of water (kg)	Weight of dry aggregate (kg)
Gravel 5-12	426	0.98	4.2	422
Sand 0-8	588	1.69	9.9	578
Sand 0-5	591	1.54	9.1	582
Total	1606	-----	23.2	1582

Table 5.102: Weight of dry aggregates for C70 company (A) cement

Aggregates Type	SSD weight (kg)	Water absorption (%)	Amount of water (kg)	Weight of dry aggregate (kg)
Gravel 5-12	412	0.98	4.0	408
Sand 0-8	568	1.69	9.6	558
Sand 0-5	571	1.54	8.8	562
Total	1552	-----	22.4	1529

Table 5.103: Weight of dry aggregates for C70 company (B) cement

Aggregates Type	SSD weight (kg)	Water absorption (%)	Amount of water (kg)	Weight of dry aggregate (kg)
Gravel 5-12	418	0.98	4.1	414
Sand 0-8	577	1.69	9.7	567
Sand 0-5	581	1.54	8.9	572
Total	1576	-----	22.7	1553

5.3.11 Final mix design sheets

Now we can complete the mix design sheets for each kind of concrete according to table 4.14. In this table, you can find all the specifications that we need for each concrete.

The only difference of tables 5.104 to 5.143 with table 4.14 is in the last section. Here we have four parts for the final mix design. Part one is the mix design of concrete for 1 m^3 and with SSD aggregates. Part two is the mix design of concrete for 1 m^3 and dry aggregates. Part three is the concrete mix design for one laboratory batch with 30 L and dry aggregates and part four is the concrete mix design for one batch of the batching plant and SSD aggregates. We use a 0.8 m^3 batch for company (A) cement and a 1.8 m^3 batch for company (B) cement.

For laboratory trials, we use dried aggregates for best accuracy. But in reality, we have all kinds of coarse aggregates dry in the batching plant and we have two wet sands. It means that they emit moisture to our concrete. But unfortunately, we don't have automatic calculator for the moisture and total water in these batching instruments. So, we make concrete in the batching plant on slump base. It means that we check the target slumps in the lab trials. Then we define the targets for the batching plant and we make concretes with the defined slump target without attention to the amount of water. In this case, we tested that the amount of water automatically is the same as we want. Because the amount of all other constituent materials is the same as the exact mix design.

Table 5.104: **The mix design for C25.1 concrete with company (A) cement**				
Batching Plant Name: **Construction Company**				
Structural Compressive Strength (MPa): **25**				
Standard Deviation (MPa): **2.5**				
Mix design Compressive Strength (MPa): **30**				
Aggregate Type	Producer	Percentage of Usage (%)	Water Absorption (%)	Water Content (%)
12-25 mm	-----	27	0.96	-----
5-12 mm	-----	15	0.98	-----
Sand 0-8	-----	29	1.69	-----
Sand 0-5	-----	29	1.54	-----
Fineness Module of Total Aggregates: **4.92**				
Type and Producer of Cement: **Company (A) ASTM Type II**				
Compressive Strength of Cement (kg/cm2): **470**				
Other Binders Type: -----				
Percent of Usage for Other Binder (%): -----				
w/b: **0.6**		Free Water (L): **182**		
Structural Element: **Foundation, Deck**				
Target Slump: **190 mm**				
Type of Super-Plasticizer: **Polycarboxylate ether BPC-40**				
Dosage of Super-Plasticizer (%): **0.3**		Water Reducing Rate (%): **15**		
Final Mix				
Constituent Material	**Weight for 1m3 (SSD)**	**Weight for 1m3 (Dry)**	**Weight for lab trial 30L (Dry)**	**Weight for one batch 0.8m3 (SSD)**
Aggregate 12-25	535 kg	530 kg	15.90 kg	428 kg
Aggregate 5-12	296 kg	293 kg	8.79 kg	237 kg
Sand 0-8	555 kg	546 kg	16.38 kg	444 kg
Sand 0-5	558 kg	549 kg	16.47 kg	447 kg
Cement	303 kg	303 kg	9.09 kg	242 kg
Other Binder	-----	-----	-----	-----
Water	182 kg	208 kg	6.24 kg	146 kg
Super-plasticizer 0.3%	0.91 kg	0.91 kg	27.3 gr	0.73 kg
Other	-----	-----	-----	-----
Total	2430 kg	2430 kg	72.94 kg	1945 kg
Other Descriptions:				

Table 5.105: **The mix design for C25.1 concrete with company (B) cement**

Batching Plant Name: **Construction Company**
Structural Compressive Strength (MPa): **25**
Standard Deviation (MPa): **2.5**
Mix design Compressive Strength (MPa): **30**

Aggregate Type	Producer	Percentage of Usage (%)	Water Absorption (%)	Water Content (%)
12-25 mm	-----	27	0.96	-----
5-12 mm	-----	15	0.98	-----
Sand 0-8	-----	29	1.69	-----
Sand 0-5	-----	29	1.54	-----

Fineness Module of Total Aggregates: **4.92**
Type and Producer of Cement: **Company (B) ASTM Type I-525**
Compressive Strength of Cement (kg/cm2): **550**
Other Binders Type: -----
Percent of Usage for Other Binder (%): -----

w/b: **0.63**	Free Water (L): **184**

Structural Element: **Foundation, Deck**
Target Slump: **190 mm**
Type of Super-Plasticizer: **Polycarboxylate ether BPC-40**

Dosage of Super-Plasticizer (%): **0.3**	Water Reducing Rate (%): **14**

Final Mix

Constituent Material	Weight for 1m3 (SSD)	Weight for 1m3 (Dry)	Weight for lab trial 30L (Dry)	Weight for one batch 1.8m3 (SSD)
Aggregate 12-25	537 kg	532 kg	15.96 kg	967 kg
Aggregate 5-12	297 kg	294 kg	8.82 kg	535 kg
Sand 0-8	557 kg	548 kg	16.44 kg	1003 kg
Sand 0-5	560 kg	551 kg	16.53 kg	1008 kg
Cement	292 kg	292 kg	8.76 kg	526 kg
Other Binder	-----	-----	-----	-----
Water	184 kg	210 kg	6.3 kg	331 kg
Super-plasticizer 0.3%	0.88 kg	0.88 kg	26.4 gr	1.58 kg
Other	-----	-----	-----	-----
Total	2428 kg	2428 kg	72.84 kg	4372 kg

Other Descriptions:

Table 5.106: **The mix design for C25.2 concrete with company (A) cement**				
Batching Plant Name: **Construction Company**				
Structural Compressive Strength (MPa): **25**				
Standard Deviation (MPa): **2.5**				
Mix design Compressive Strength (MPa): **30**				
Aggregate Type	Producer	Percentage of Usage (%)	Water Absorption (%)	Water Content (%)
11-19 mm	-----	23	0.98	-----
5-12 mm	-----	11	0.98	-----
Sand 0-8	-----	33	1.69	-----
Sand 0-5	-----	33	1.54	-----
Fineness Module of Total Aggregates: **4.54**				
Type and Producer of Cement: **Company (B) ASTM Type II**				
Compressive Strength of Cement (kg/cm2): **470**				
Other Binders Type: -----				
Percent of Usage for Other Binder (%): -----				
w/b: **0.6**		Free Water (L): **194**		
Structural Element: **Foundation, Deck, Column, Wall**				
Target Slump: **190 mm**				
Type of Super-Plasticizer: **Polycarboxylate ether BPC-40**				
Dosage of Super-Plasticizer (%): **0.3**		Water Reducing Rate (%): **15**		

Final Mix				
Constituent Material	**Weight for 1m3 (SSD)**	**Weight for 1m3 (Dry)**	**Weight for lab trial 30L (Dry)**	**Weight for one batch 0.8m3 (SSD)**
Aggregate 11-19	442 kg	438 kg	13.14 kg	354 kg
Aggregate 5-12	211 kg	209 kg	6.27 kg	169 kg
Sand 0-8	613 kg	603 kg	18.09 kg	490 kg
Sand 0-5	616 kg	607 kg	18.21 kg	493 kg
Cement	323 kg	323 kg	9.69 kg	258 kg
Other Binder	-----	-----	-----	-----
Water	194 kg	220 kg	6.6 kg	155 kg
Super-plasticizer 0.3%	0.97 kg	0.97 kg	29.1 gr	0.78 kg
Other	-----	-----	-----	-----
Total	2400 kg	2400 kg	72.03 kg	1920 kg
Other Descriptions:				

Table 5.107: **The mix design for C25.2 concrete with company (B) cement**				
Batching Plant Name: **Construction Company**				
Structural Compressive Strength (MPa): **25**				
Standard Deviation (MPa): **2.5**				
Mix design Compressive Strength (MPa): **30**				
Aggregate Type	Producer	Percentage of Usage (%)	Water Absorption (%)	Water Content (%)
11-19 mm	-----	23	0.98	-----
5-12 mm	-----	11	0.98	-----
Sand 0-8	-----	33	1.69	-----
Sand 0-5	-----	33	1.54	-----
Fineness Module of Total Aggregates: **4.54**				
Type and Producer of Cement: **Company (B) ASTM Type I-525**				
Compressive Strength of Cement (kg/cm2): **550**				
Other Binders Type: -----				
Percent of Usage for Other Binder (%): -----				
w/b: **0.63**		Free Water (L): **196**		
Structural Element: **Foundation, Deck, Column, Wall**				
Target Slump: **190 mm**				
Type of Super-Plasticizer: **Polycarboxylate ether BPC-40**				
Dosage of Super-Plasticizer (%): **0.3**		Water Reducing Rate (%): **14**		
Final Mix				
Constituent Material	**Weight for 1m3 (SSD)**	**Weight for 1m3 (Dry)**	**Weight for lab trial 30L (Dry)**	**Weight for one batch 1.8m3 (SSD)**
Aggregate 11-19	443 kg	439 kg	13.17 kg	797 kg
Aggregate 5-12	211 kg	209 kg	6.27 kg	380 kg
Sand 0-8	615 kg	605 kg	18.15 kg	1107 kg
Sand 0-5	618 kg	608 kg	18.24 kg	1112 kg
Cement	311 kg	311 kg	9.33 kg	560 kg
Other Binder	-----	-----	-----	-----
Water	196 kg	222 kg	6.66 kg	353 kg
Super-plasticizer 0.3%	0.93 kg	0.93 kg	27.9 gr	1.67 kg
Other	-----	-----	-----	-----
Total	2395 kg	2395 kg	71.85 kg	4311 kg
Other Descriptions:				

Table 5.108: **The mix design for C30.1 concrete with company (A) cement**					
Batching Plant Name: **Construction Company**					
Structural Compressive Strength (MPa): **30**					
Standard Deviation (MPa): **2.5**					
Mix design Compressive Strength (MPa): **35**					
Aggregate Type	Producer	Percentage of Usage (%)	Water Absorption (%)	Water Content (%)	
11-19 mm	-----	23	0.98	-----	
5-12 mm	-----	11	0.98	-----	
Sand 0-8	-----	33	1.69	-----	
Sand 0-5	-----	33	1.54	-----	
Fineness Module of Total Aggregates: **4.54**					
Type and Producer of Cement: **Company (A) ASTM Type II**					
Compressive Strength of Cement (kg/cm2): **470**					
Other Binders Type: -----					
Percent of Usage for Other Binder (%): -----					
w/b: **0.54**			Free Water (L): **189**		
Structural Element: **Foundation, Deck, Column, Wall**					
Target Slump: **190 mm**					
Type of Super-Plasticizer: **Polycarboxylate ether BPC-40**					
Dosage of Super-Plasticizer (%): **0.5**			Water Reducing Rate (%): **19**		

Final Mix				
Constituent Material	**Weight for 1m3 (SSD)**	**Weight for 1m3 (Dry)**	**Weight for lab trial 30L (Dry)**	**Weight for one batch 0.8m3 (SSD)**
Aggregate 11-19	439 kg	435 kg	13.05 kg	351 kg
Aggregate 5-12	210 kg	208 kg	6.24 kg	168 kg
Sand 0-8	609 kg	599 kg	17.97 kg	487 kg
Sand 0-5	613 kg	604 kg	18.12 kg	490 kg
Cement	350 kg	350 kg	10.50 kg	280 kg
Other Binder	-----	-----	-----	-----
Water	189 kg	215 kg	6.45 kg	151 kg
Super-plasticizer 0.5%	1.75 kg	1.75 kg	52.5 gr	1.4 kg
Other	-----	-----	-----	-----
Total	2412 kg	2412 kg	72.38 kg	1928 kg
Other Descriptions:				

Table 5.109: **The mix design for C30.1 concrete with company (B) cement**				
Batching Plant Name: **Construction Company**				
Structural Compressive Strength (MPa): **30**				
Standard Deviation (MPa): **2.5**				
Mix design Compressive Strength (MPa): **35**				
Aggregate Type	Producer	Percentage of Usage (%)	Water Absorption (%)	Water Content (%)
11-19 mm	-----	23	0.98	-----
5-12 mm	-----	11	0.98	-----
Sand 0-8	-----	33	1.69	-----
Sand 0-5	-----	33	1.54	-----
Fineness Module of Total Aggregates: **4.54**				
Type and Producer of Cement: **Company (B) ASTM Type I-525**				
Compressive Strength of Cement (kg/cm2): **550**				
Other Binders Type: -----				
Percent of Usage for Other Binder (%): -----				
w/b: **0.57**		Free Water (L): **192**		
Structural Element: **Foundation, Deck, Column, Wall**				
Target Slump: **190 mm**				
Type of Super-Plasticizer: **Polycarboxylate ether BPC-40**				
Dosage of Super-Plasticizer (%): **0.5**		Water Reducing Rate (%): **18**		

Final Mix				
Constituent Material	**Weight for 1m3 (SSD)**	**Weight for 1m3 (Dry)**	**Weight for lab trial 30L (Dry)**	**Weight for one batch 1.8m3 (SSD)**
Aggregate 11-19	439 kg	435 kg	13.05 kg	790 kg
Aggregate 5-12	210 kg	208 kg	6.24 kg	378 kg
Sand 0-8	609 kg	599 kg	17.97 kg	1096 kg
Sand 0-5	613 kg	604 kg	18.12 kg	1103 kg
Cement	337 kg	337 kg	10.11 kg	607 kg
Other Binder	-----	-----	-----	-----
Water	192 kg	218 kg	6.54 kg	346 kg
Super-plasticizer 0.5%	1.69 kg	1.69 kg	50.7 gr	3.04 kg
Other	-----	-----	-----	-----
Total	2402 kg	2402 kg	72.08 kg	4323 kg
Other Descriptions:				

Table 5.110: **The mix design for C30.2 concrete with company (A) cement**

Batching Plant Name: **Construction Company**
Structural Compressive Strength (MPa): **30**
Standard Deviation (MPa): **2.5**
Mix design Compressive Strength (MPa): **35**

Aggregate Type	Producer	Percentage of Usage (%)	Water Absorption (%)	Water Content (%)
5-12 mm	-----	26	0.98	-----
Sand 0-8	-----	37	1.69	-----
Sand 0-5	-----	37	1.54	-----

Fineness Module of Total Aggregates: **4.19**	
Type and Producer of Cement: **Company (A) ASTM Type II**	
Compressive Strength of Cement (kg/cm2): **470**	
Other Binders Type: -----	
Percent of Usage for Other Binder (%): -----	
w/b: **0.54**	Free Water (L): **199**
Structural Element: **Deck, Column, Wall**	
Target Slump: **190 mm**	
Type of Super-Plasticizer: **Polycarboxylate ether BPC-40**	
Dosage of Super-Plasticizer (%): **0.5**	Water Reducing Rate (%): **19**

Final Mix

Constituent Material	Weight for 1m3 (SSD)	Weight for 1m3 (Dry)	Weight for lab trial 30L (Dry)	Weight for one batch 0.8m3 (SSD)
Aggregate 5-12	482 kg	477 kg	14.31 kg	386 kg
Sand 0-8	664 kg	653 kg	19.59 kg	531 kg
Sand 0-5	668 kg	658 kg	19.74 kg	534 kg
Cement	368 kg	368 kg	11.04 kg	295 kg
Other Binder	-----	-----	-----	-----
Water	199 kg	225 kg	6.75 kg	159 kg
Super-plasticizer 0.5%	1.84 kg	1.84 kg	55.2 gr	1.47 kg
Other	-----	-----	-----	-----
Total	2383 kg	2383 kg	71.48 kg	1906 kg

Other Descriptions:

Table 5.111: **The mix design for C30.2 concrete with company (B) cement**

Batching Plant Name: **Construction Company**

Structural Compressive Strength (MPa): 30

Standard Deviation (MPa): **2.5**

Mix design Compressive Strength (MPa): **35**

Aggregate Type	Producer	Percentage of Usage (%)	Water Absorption (%)	Water Content (%)
5-12 mm	-----	26	0.98	-----
Sand 0-8	-----	37	1.69	-----
Sand 0-5	-----	37	1.54	-----

Fineness Module of Total Aggregates: **4.19**

Type and Producer of Cement: **Company (B) ASTM Type I-525**

Compressive Strength of Cement (kg/cm2): **550**

Other Binders Type: -----

Percent of Usage for Other Binder (%): -----

w/b: **0.57**	Free Water (L): **202**

Structural Element: **Deck, Column, Wall**

Target Slump: **190 mm**

Type of Super-Plasticizer: **Polycarboxylate ether BPC-40**

Dosage of Super-Plasticizer (%): **0.5**	Water Reducing Rate (%): **18**

Final Mix

Constituent Material	Weight for 1m3 (SSD)	Weight for 1m3 (Dry)	Weight for lab trial 30L (Dry)	Weight for one batch 1.8m3 (SSD)
Aggregate 5-12	482 kg	477 kg	14.31 kg	868 kg
Sand 0-8	664 kg	653 kg	19.59 kg	1195 kg
Sand 0-5	668 kg	658 kg	19.74 kg	1202 kg
Cement	355 kg	355 kg	10.65 kg	639 kg
Other Binder	-----	-----	-----	-----
Water	202 kg	228 kg	6.84 kg	364 kg
Super-plasticizer 0.5%	1.78 kg	1.78 kg	53.4 gr	3.2 kg
Other	-----	-----	-----	-----
Total	2373 kg	2373 kg	71.18 kg	4271 kg

Other Descriptions:

Table 5.112: **The mix design for C30.3 concrete with company (A) cement**				
Batching Plant Name: **Construction Company**				
Structural Compressive Strength (MPa): **30**				
Standard Deviation (MPa): **2.5**				
Mix design Compressive Strength (MPa): **35**				
Aggregate Type	Producer	Percentage of Usage (%)	Water Absorption (%)	Water Content (%)
11-19 mm	-----	23	0.98	-----
5-12 mm	-----	11	0.98	-----
Sand 0-8	-----	33	1.69	-----
Sand 0-5	-----	33	1.54	-----
Fineness Module of Total Aggregates: **4.54**				
Type and Producer of Cement: **Company (A) ASTM Type II**				
Compressive Strength of Cement (kg/cm2): **470**				
Other Binders Type: **GGBS**				
Percent of Usage for Other Binder (%): **20**				
w/b: **0.54**		Free Water (L): **189**		
Structural Element: **Foundation, Deck, Column, Wall**				
Target Slump: **190 mm**				
Type of Super-Plasticizer: **Polycarboxylate ether BPC-40**				
Dosage of Super-Plasticizer (%): **0.5**		Water Reducing Rate (%): **19**		
Final Mix				
Constituent Material	**Weight for 1m3 (SSD)**	**Weight for 1m3 (Dry)**	**Weight for lab trial 30L (Dry)**	**Weight for one batch 0.8m3 (SSD)**
Aggregate 11-19	**427 kg**	**423 kg**	**12.69 kg**	**342 kg**
Aggregate 5-12	**204 kg**	**202 kg**	**6.06 kg**	**163 kg**
Sand 0-8	**592 kg**	**582 kg**	**17.46 kg**	**474 kg**
Sand 0-5	**596 kg**	**587 kg**	**17.61 kg**	**477 kg**
Cement	**322 kg**	**322 kg**	**9.66 kg**	**258 kg**
Other Binder (GGBS)	**80 kg**	**80 kg**	**2.40 kg**	**64 kg**
Water	**189 kg**	**214 kg**	**6.42 kg**	**151 kg**
Super-plasticizer 0.5%	**2.01 kg**	**2.01 kg**	**60.6 gr**	**1.61 kg**
Other	-----	-----	-----	-----
Total	**2412 kg**	**2412 kg**	**72.36 kg**	**1931 kg**
Other Descriptions:				

Table 5.113: **The mix design for C30.3 concrete with company (B) cement**				
Batching Plant Name: **Construction Company**				
Structural Compressive Strength (MPa): **30**				
Standard Deviation (MPa): **2.5**				
Mix design Compressive Strength (MPa): **35**				
Aggregate Type	Producer	Percentage of Usage (%)	Water Absorption (%)	Water Content (%)
11-19 mm	-----	23	0.98	-----
5-12 mm	-----	11	0.98	-----
Sand 0-8	-----	33	1.69	-----
Sand 0-5	-----	33	1.54	-----
Fineness Module of Total Aggregates: **4.54**				
Type and Producer of Cement: **Company (B) ASTM Type I-525**				
Compressive Strength of Cement (kg/cm2): **550**				
Other Binders Type: **GGBS**				
Percent of Usage for Other Binder (%): **20**				
w/b: **0.57**		Free Water (L): **192**		
Structural Element: **Foundation, Deck, Column, Wall**				
Target Slump: **190 mm**				
Type of Super-Plasticizer: **Polycarboxylate ether BPC-40**				
Dosage of Super-Plasticizer (%): **0.5**		Water Reducing Rate (%): **18**		

Final Mix

Constituent Material	Weight for 1m3 (SSD)	Weight for 1m3 (Dry)	Weight for lab trial 30L (Dry)	Weight for one batch 1.8m3 (SSD)
Aggregate 11-19	428 kg	424 kg	12.72 kg	770 kg
Aggregate 5-12	204 kg	202 kg	6.06 kg	367 kg
Sand 0-8	593 kg	583 kg	17.49 kg	1067 kg
Sand 0-5	596 kg	587 kg	17.61 kg	1073 kg
Cement	310 kg	310 kg	9.30 kg	558 kg
Other Binder (GGBS)	78 kg	78 kg	2.34 kg	140 kg
Water	192 kg	217 kg	6.51 kg	346 kg
Super-plasticizer 0.5%	1.94 kg	1.94 kg	58.2 gr	3.49 kg
Other	-----	-----	-----	-----
Total	2403 kg	2403 kg	72.09 kg	4324 kg
Other Descriptions:				

Table 5.114: **The mix design for C30.4 concrete with company (A) cement**				
Batching Plant Name: **Construction Company**				
Structural Compressive Strength (MPa): **30**				
Standard Deviation (MPa): **2.5**				
Mix design Compressive Strength (MPa): **35**				
Aggregate Type	Producer	Percentage of Usage (%)	Water Absorption (%)	Water Content (%)
5-12 mm	-----	26	0.98	-----
Sand 0-8	-----	37	1.69	-----
Sand 0-5	-----	37	1.54	-----
Fineness Module of Total Aggregates: **4.19**				
Type and Producer of Cement: **Company (A) ASTM Type II**				
Compressive Strength of Cement (kg/cm2): **470**				
Other Binders Type: **GGBS**				
Percent of Usage for Other Binder (%): **20**				
w/b: **0.54**		Free Water (L): **199**		
Structural Element: **Deck, Column, Wall**				
Target Slump: **190 mm**				
Type of Super-Plasticizer: **Polycarboxylate ether BPC-40**				
Dosage of Super-Plasticizer (%): **0.5**		Water Reducing Rate (%): **19**		

Final Mix

Constituent Material	Weight for 1m3 (SSD)	Weight for 1m3 (Dry)	Weight for lab trial 30L (Dry)	Weight for one batch 0.8m3 (SSD)
Aggregate 5-12	467 kg	462 kg	13.86 kg	374 kg
Sand 0-8	644 kg	633 kg	18.99 kg	515 kg
Sand 0-5	648 kg	638 kg	19.14 kg	518 kg
Cement	338 kg	338 kg	10.14 kg	270 kg
Other Binder (GGBS)	85 kg	85 kg	2.55 kg	68 kg
Water	199 kg	224 kg	6.72 kg	159 kg
Super-plasticizer 0.5%	2.11 kg	2.11 kg	63.3 gr	1.69 kg
Other	-----	-----	-----	-----
Total	2383 kg	2382 kg	71.46 kg	1906 kg
Other Descriptions:				

Table 5.115: **The mix design for C30.4 concrete with company (B) cement**				
Batching Plant Name: **Construction Company**				
Structural Compressive Strength (MPa): **30**				
Standard Deviation (MPa): **2.5**				
Mix design Compressive Strength (MPa): **35**				
Aggregate Type	Producer	Percentage of Usage (%)	Water Absorption (%)	Water Content (%)
5-12 mm	-----	26	0.98	-----
Sand 0-8	-----	37	1.69	-----
Sand 0-5	-----	37	1.54	-----
Fineness Module of Total Aggregates: **4.19**				
Type and Producer of Cement: **Company (B) ASTM Type I-525**				
Compressive Strength of Cement (kg/cm2): **550**				
Other Binders Type: **GGBS**				
Percent of Usage for Other Binder (%): **20**				
w/b: **0.57**		Free Water (L): **202**		
Structural Element: **Deck, Column, Wall**				
Target Slump: **190 mm**				
Type of Super-Plasticizer: **Polycarboxylate ether BPC-40**				
Dosage of Super-Plasticizer (%): **0.5**		Water Reducing Rate (%): **18**		

Final Mix

Constituent Material	Weight for 1m3 (SSD)	Weight for 1m3 (Dry)	Weight for lab trial 30L (Dry)	Weight for one batch 1.8m3 (SSD)
Aggregate 5-12	468 kg	463 kg	13.89 kg	842 kg
Sand 0-8	645 kg	634 kg	19.02 kg	1161 kg
Sand 0-5	649 kg	639 kg	19.17 kg	1168 kg
Cement	326 kg	326 kg	9.78 kg	587 kg
Other Binder (GGBS)	82 kg	82 kg	2.46 kg	148 kg
Water	202 kg	227 kg	6.81 kg	364 kg
Super-plasticizer 0.5%	2.04 kg	2.04 kg	61.2 gr	3.67 kg
Other	-----	-----	-----	-----
Total	2374 kg	2373 kg	71.19 kg	4273 kg
Other Descriptions:				

Table 5.116: **The mix design for C35.1 concrete with company (A) cement**

Batching Plant Name: **Construction Company**

Structural Compressive Strength (MPa): **35**

Standard Deviation (MPa): **2.5**

Mix design Compressive Strength (MPa): **40**

Aggregate Type	Producer	Percentage of Usage (%)	Water Absorption (%)	Water Content (%)
11-19 mm	-----	23	0.98	-----
5-12 mm	-----	11	0.98	-----
Sand 0-8	-----	33	1.69	-----
Sand 0-5	-----	33	1.54	-----

Fineness Module of Total Aggregates: **4.54**

Type and Producer of Cement: **Company (A) ASTM Type II**

Compressive Strength of Cement (kg/cm2): **470**

Other Binders Type: -----

Percent of Usage for Other Binder (%): -----

w/b: **0.49**	Free Water (L): **188**

Structural Element: **Foundation, Deck, Column, Wall**

Target Slump: **190 mm**

Type of Super-Plasticizer: **Polycarboxylate ether BPC-40**

Dosage of Super-Plasticizer (%): **0.6**	Water Reducing Rate (%): **21**

Final Mix

Constituent Material	Weight for 1m3 (SSD)	Weight for 1m3 (Dry)	Weight for lab trial 30L (Dry)	Weight for one batch 0.8m3 (SSD)
Aggregate 11-19	432 kg	428 kg	12.84 kg	346 kg
Aggregate 5-12	206 kg	204 kg	6.12 kg	165 kg
Sand 0-8	599 kg	589 kg	17.67 kg	479 kg
Sand 0-5	603 kg	594 kg	17.82 kg	482 kg
Cement	384 kg	384 kg	11.52 kg	307 kg
Other Binder	-----	-----	-----	-----
Water	188 kg	214 kg	6.42 kg	150 kg
Super-plasticizer 0.6%	2.30 kg	2.30 kg	69.0 gr	1.84 kg
Other	-----	-----	-----	-----
Total	2414 kg	2415 kg	72.46 kg	1931 kg

Other Descriptions:

Table 5.117: **The mix design for C35.1 concrete with company (B) cement**				
Batching Plant Name: **Construction Company**				
Structural Compressive Strength (MPa): **35**				
Standard Deviation (MPa): **2.5**				
Mix design Compressive Strength (MPa): **40**				
Aggregate Type	Producer	Percentage of Usage (%)	Water Absorption (%)	Water Content (%)
11-19 mm	-----	23	0.98	-----
5-12 mm	-----	11	0.98	-----
Sand 0-8	-----	33	1.69	-----
Sand 0-5	-----	33	1.54	-----
Fineness Module of Total Aggregates: **4.54**				
Type and Producer of Cement: **Company (B) ASTM Type I-525**				
Compressive Strength of Cement (kg/cm2): **550**				
Other Binders Type: -----				
Percent of Usage for Other Binder (%): -----				
w/b: **0.5**		Free Water (L): **190**		
Structural Element: **Foundation, Deck, Column, Wall**				
Target Slump: **190 mm**				
Type of Super-Plasticizer: **Polycarboxylate ether BPC-40**				
Dosage of Super-Plasticizer (%): **0.6**		Water Reducing Rate (%): **20**		
Final Mix				
Constituent Material	**Weight for 1m3 (SSD)**	**Weight for 1m3 (Dry)**	**Weight for lab trial 30L (Dry)**	**Weight for one batch 1.8m3 (SSD)**
Aggregate 11-19	431 kg	427 kg	12.81 kg	776 kg
Aggregate 5-12	206 kg	204 kg	6.12 kg	371 kg
Sand 0-8	599 kg	589 kg	17.67 kg	1078 kg
Sand 0-5	602 kg	593 kg	17.79 kg	1084 kg
Cement	380 kg	380 kg	11.40 kg	684 kg
Other Binder	-----	-----	-----	-----
Water	190 kg	216 kg	6.48 kg	342 kg
Super-plasticizer 0.6%	2.28 kg	2.28 kg	68.4 gr	4.10 kg
Other	-----	-----	-----	-----
Total	2410 kg	2411 kg	72.34 kg	4339 kg
Other Descriptions:				

Table 5.118: **The mix design for C35.2 concrete with company (A) cement**				
Batching Plant Name: **Construction Company**				
Structural Compressive Strength (MPa): **35**				
Standard Deviation (MPa): **2.5**				
Mix design Compressive Strength (MPa): **40**				
Aggregate Type	Producer	Percentage of Usage (%)	Water Absorption (%)	Water Content (%)
5-12 mm	-----	26	0.98	-----
Sand 0-8	-----	37	1.69	-----
Sand 0-5	-----	37	1.54	-----
Fineness Module of Total Aggregates: **4.19**				
Type and Producer of Cement: **Company (A) ASTM Type II**				
Compressive Strength of Cement (kg/cm2): **470**				
Other Binders Type: -----				
Percent of Usage for Other Binder (%): -----				
w/b: **0.49**		Free Water (L): **197**		
Structural Element: **Deck, Column, Wall**				
Target Slump: **190 mm**				
Type of Super-Plasticizer: **Polycarboxylate ether BPC-40**				
Dosage of Super-Plasticizer (%): **0.6**		Water Reducing Rate (%): **21**		

Final Mix

Constituent Material	Weight for 1m3 (SSD)	Weight for 1m3 (Dry)	Weight for lab trial 30L (Dry)	Weight for one batch 0.8m3 (SSD)
Aggregate 5-12	475 kg	470 kg	14.10 kg	380 kg
Sand 0-8	655 kg	644 kg	19.32 kg	524 kg
Sand 0-5	659 kg	649 kg	19.47 kg	527 kg
Cement	402 kg	402 kg	12.06 kg	322 kg
Other Binder	-----	-----	-----	-----
Water	197 kg	223 kg	6.69 kg	158 kg
Super-plasticizer 0.6%	2.41 kg	2.41 kg	72.3 gr	1.93 kg
Other	-----	-----	-----	-----
Total	2390 kg	2390 kg	71.71 kg	1913 kg
Other Descriptions:				

Table 5.119: **The mix design for C35.2 concrete with company (B) cement**				
Batching Plant Name: **Construction Company**				
Structural Compressive Strength (MPa): **35**				
Standard Deviation (MPa): **2.5**				
Mix design Compressive Strength (MPa): **40**				
Aggregate Type	Producer	Percentage of Usage (%)	Water Absorption (%)	Water Content (%)
5-12 mm	-----	26	0.98	-----
Sand 0-8	-----	37	1.69	-----
Sand 0-5	-----	37	1.54	-----
Fineness Module of Total Aggregates: **4.19**				
Type and Producer of Cement: **Company (B) ASTM Type I-525**				
Compressive Strength of Cement (kg/cm2): **550**				
Other Binders Type: -----				
Percent of Usage for Other Binder (%): -----				
w/b: **0.5**		Free Water (L): **200**		
Structural Element: **Deck, Column, Wall**				
Target Slump: **190 mm**				
Type of Super-Plasticizer: **Polycarboxylate ether BPC-40**				
Dosage of Super-Plasticizer (%): **0.6**		Water Reducing Rate (%): **20**		

	Final Mix			
Constituent Material	**Weight for 1m3 (SSD)**	**Weight for 1m3 (Dry)**	**Weight for lab trial 30L (Dry)**	**Weight for one batch 1.8m3 (SSD)**
Aggregate 5-12	473 kg	468 kg	14.04 kg	851 kg
Sand 0-8	652 kg	641 kg	19.23 kg	1174 kg
Sand 0-5	656 kg	646 kg	19.38 kg	1181 kg
Cement	400 kg	400 kg	12.00 kg	720 kg
Other Binder	-----	-----	-----	-----
Water	200 kg	226 kg	6.78 kg	360 kg
Super-plasticizer 0.6%	2.40 kg	2.40 kg	72.0 gr	4.32 kg
Other	-----	-----	-----	-----
Total	2383 kg	2383 kg	71.50 kg	4290 kg
Other Descriptions:				

Table 5.120: **The mix design for C35.3 concrete with company (A) cement**				
Batching Plant Name: **Construction Company**				
Structural Compressive Strength (MPa): **35**				
Standard Deviation (MPa): **2.5**				
Mix design Compressive Strength (MPa): **40**				
Aggregate Type	Producer	Percentage of Usage (%)	Water Absorption (%)	Water Content (%)
11-19 mm	-----	23	0.98	-----
5-12 mm	-----	11	0.98	-----
Sand 0-8	-----	33	1.69	-----
Sand 0-5	-----	33	1.54	-----
Fineness Module of Total Aggregates: **4.54**				
Type and Producer of Cement: **Company (A) ASTM Type II**				
Compressive Strength of Cement (kg/cm2): **470**				
Other Binders Type: **GGBS**				
Percent of Usage for Other Binder (%): **20**				
w/b: **0.49**		Free Water (L): **188**		
Structural Element: **Foundation, Deck, Column, Wall**				
Target Slump: **190 mm**				
Type of Super-Plasticizer: **Polycarboxylate ether BPC-40**				
Dosage of Super-Plasticizer (%): **0.6**		Water Reducing Rate (%): **21**		
Final Mix				
Constituent Material	**Weight for 1m3 (SSD)**	**Weight for 1m3 (Dry)**	**Weight for lab trial 30L (Dry)**	**Weight for one batch 0.8m3 (SSD)**
Aggregate 11-19	419 kg	415 kg	12.45 kg	335 kg
Aggregate 5-12	200 kg	198 kg	5.94 kg	160 kg
Sand 0-8	582 kg	572 kg	17.16 kg	466 kg
Sand 0-5	585 kg	576 kg	17.28 kg	468 kg
Cement	354 kg	354 kg	10.62 kg	283 kg
Other Binder (GGBS)	88 kg	88 kg	2.64 kg	70 kg
Water	188 kg	213 kg	6.39 kg	150 kg
Super-plasticizer 0.6%	2.65 kg	2.65 kg	79.5 gr	2.12 kg
Other	-----	-----	-----	-----
Total	2418 kg	2418 kg	72.56 kg	1934 kg
Other Descriptions:				

Table 5.121: **The mix design for C35.3 concrete with company (B) cement**				
Batching Plant Name: **Construction Company**				
Structural Compressive Strength (MPa): **35**				
Standard Deviation (MPa): **2.5**				
Mix design Compressive Strength (MPa): **40**				
Aggregate Type	Producer	Percentage of Usage (%)	Water Absorption (%)	Water Content (%)
11-19 mm	-----	23	0.98	-----
5-12 mm	-----	11	0.98	-----
Sand 0-8	-----	33	1.69	-----
Sand 0-5	-----	33	1.54	-----
Fineness Module of Total Aggregates: **4.54**				
Type and Producer of Cement: **Company (B) ASTM Type I-525**				
Compressive Strength of Cement (kg/cm2): **550**				
Other Binders Type: **GGBS**				
Percent of Usage for Other Binder (%): **20**				
w/b: **0.50**		Free Water (L): **190**		
Structural Element: **Foundation, Deck, Column, Wall**				
Target Slump: **190 mm**				
Type of Super-Plasticizer: **Polycarboxylate ether BPC-40**				
Dosage of Super-Plasticizer (%): **0.6**		Water Reducing Rate (%): **20**		
Final Mix				
Constituent Material	**Weight for 1m3 (SSD)**	**Weight for 1m3 (Dry)**	**Weight for lab trial 30L (Dry)**	**Weight for one batch 1.8m3 (SSD)**
Aggregate 11-19	419 kg	415 kg	12.45 kg	754 kg
Aggregate 5-12	200 kg	198 kg	5.94 kg	360 kg
Sand 0-8	581 kg	571 kg	17.13 kg	1046 kg
Sand 0-5	584 kg	576 kg	17.28 kg	1051 kg
Cement	350 kg	350 kg	10.50 kg	630 kg
Other Binder (GGBS)	87 kg	87 kg	2.61 kg	157 kg
Water	190 kg	215 kg	6.45 kg	342 kg
Super-plasticizer 0.6%	2.62 kg	2.62 kg	78.6 gr	4.72 kg
Other	-----	-----	-----	-----
Total	2413 kg	2414 kg	72.44 kg	4345 kg
Other Descriptions:				

Table 5.122: **The mix design for C35.4 concrete with company (A) cement**				
Batching Plant Name: **Construction Company**				
Structural Compressive Strength (MPa): **35**				
Standard Deviation (MPa): **2.5**				
Mix design Compressive Strength (MPa): **40**				
Aggregate Type	Producer	Percentage of Usage (%)	Water Absorption (%)	Water Content (%)
5-12 mm	-----	26	0.98	-----
Sand 0-8	-----	37	1.69	-----
Sand 0-5	-----	37	1.54	-----
Fineness Module of Total Aggregates: **4.19**				
Type and Producer of Cement: **Company (A) ASTM Type II**				
Compressive Strength of Cement (kg/cm2): **470**				
Other Binders Type: **GGBS**				
Percent of Usage for Other Binder (%): **20**				
w/b: **0.49**		Free Water (L): **197**		
Structural Element: **Deck, Column, Wall**				
Target Slump: **190 mm**				
Type of Super-Plasticizer: **Polycarboxylate ether BPC-40**				
Dosage of Super-Plasticizer (%): **0.6**		Water Reducing Rate (%): **21**		

Final Mix

Constituent Material	Weight for 1m3 (SSD)	Weight for 1m3 (Dry)	Weight for lab trial 30L (Dry)	Weight for One batch 0.8m3 (SSD)
Aggregate 5-12	459 kg	455 kg	13.65 kg	367 kg
Sand 0-8	633 kg	622 kg	18.66 kg	506 kg
Sand 0-5	637 kg	627 kg	18.81 kg	510 kg
Cement	370 kg	370 kg	11.10 kg	296 kg
Other Binder (GGBS)	92 kg	92 kg	2.76 kg	74 kg
Water	197 kg	222 kg	6.66 kg	158 kg
Super-plasticizer 0.6%	2.77 kg	2.77 kg	83.1 gr	2.22 kg
Other	-----	-----	-----	-----
Total	2391 kg	2391 kg	71.72 kg	1913 kg
Other Descriptions:				

Table 5.123: **The mix design for C35.4 concrete with company (B) cement**				
Batching Plant Name: **Construction Company**				
Structural Compressive Strength (MPa): **35**				
Standard Deviation (MPa): **2.5**				
Mix design Compressive Strength (MPa): **40**				
Aggregate Type	Producer	Percentage of Usage (%)	Water Absorption (%)	Water Content (%)
5-12 mm	-----	26	0.98	-----
Sand 0-8	-----	37	1.69	-----
Sand 0-5	-----	37	1.54	-----
Fineness Module of Total Aggregates: **4.19**				
Type and Producer of Cement: **Company (B) ASTM Type I-525**				
Compressive Strength of Cement (kg/cm2): **550**				
Other Binders Type: **GGBS**				
Percent of Usage for Other Binder (%): **20**				
w/b: **0.50**		Free Water (L): **200**		
Structural Element: **Deck, Column, Wall**				
Target Slump: **190 mm**				
Type of Super-Plasticizer: **Polycarboxylate ether BPC-40**				
Dosage of Super-Plasticizer (%): **0.6**		Water Reducing Rate (%): **20**		

Final Mix				
Constituent Material	**Weight for 1m3 (SSD)**	**Weight for 1m3 (Dry)**	**Weight for lab trial 30L (Dry)**	**Weight for one batch 1.8m3 (SSD)**
Aggregate 5-12	**457 kg**	**453 kg**	**13.59 kg**	**823 kg**
Sand 0-8	**630 kg**	**619 kg**	**18.57 kg**	**1134 kg**
Sand 0-5	**634 kg**	**624 kg**	**18.72 kg**	**1141 kg**
Cement	**368 kg**	**368 kg**	**11.04 kg**	**662 kg**
Other Binder (GGBS)	**92 kg**	**92 kg**	**2.76 kg**	**166 kg**
Water	**200 kg**	**225 kg**	**6.75 kg**	**360 kg**
Super-plasticizer 0.6%	**2.76 kg**	**2.76 kg**	**82.8 gr**	**4.97 kg**
Other	-----	-----	-----	-----
Total	**2384 kg**	**2384 kg**	**71.51 kg**	**4291 kg**
Other Descriptions:				

Table 5.124: **The mix design for C40.1 concrete with company (A) cement**				
Batching Plant Name: **Construction Company**				
Structural Compressive Strength (MPa): **40**				
Standard Deviation (MPa): **2.5**				
Mix design Compressive Strength (MPa): **45**				
Aggregate Type	Producer	Percentage of Usage (%)	Water Absorption (%)	Water Content (%)
11-19 mm	-----	23	0.98	-----
5-12 mm	-----	11	0.98	-----
Sand 0-8	-----	33	1.69	-----
Sand 0-5	-----	33	1.54	-----
Fineness Module of Total Aggregates: **4.54**				
Type and Producer of Cement: **Company (A) ASTM Type II**				
Compressive Strength of Cement (kg/cm2): **470**				
Other Binders Type: -----				
Percent of Usage for Other Binder (%): -----				
w/b: **0.43**		Free Water (L): **186**		
Structural Element: **Foundation, Deck, Column, Wall**				
Target Slump: **190 mm**				
Type of Super-Plasticizer: **Polycarboxylate ether BPC-40**				
Dosage of Super-Plasticizer (%): **0.7**		Water Reducing Rate (%): **23**		

Final Mix

Constituent Material	Weight for 1m3 (SSD)	Weight for 1m3 (Dry)	Weight for lab trial 30L (Dry)	Weight for one batch 0.8m3 (SSD)
Aggregate 11-19	**424 kg**	**420 kg**	**12.60 kg**	**339 kg**
Aggregate 5-12	**202 kg**	**200 kg**	**6.00 kg**	**162 kg**
Sand 0-8	**588 kg**	**578 kg**	**17.34 kg**	**470 kg**
Sand 0-5	**591 kg**	**582 kg**	**17.46 kg**	**473 kg**
Cement	**433 kg**	**433 kg**	**12.99 kg**	**346 kg**
Other Binder	-----	-----	-----	-----
Water	**186 kg**	**211 kg**	**6.33 kg**	**149 kg**
Super-plasticizer 0.7%	**3.03 kg**	**3.03 kg**	**90.9 gr**	**2.42 kg**
Other	-----	-----	-----	-----
Total	**2427 kg**	**2427 kg**	**72.81 kg**	**1941 kg**
Other Descriptions:				

Table 5.125: **The mix design for C40.1 concrete with company (B) cement**				
Batching Plant Name: **Construction Company**				
Structural Compressive Strength (MPa): **40**				
Standard Deviation (MPa): **2.5**				
Mix design Compressive Strength (MPa): **45**				
Aggregate Type	Producer	Percentage of Usage (%)	Water Absorption (%)	Water Content (%)
11-19 mm	-----	23	0.98	-----
5-12 mm	-----	11	0.98	-----
Sand 0-8	-----	33	1.69	-----
Sand 0-5	-----	33	1.54	-----
Fineness Module of Total Aggregates: **4.54**				
Type and Producer of Cement: **Company (B) ASTM Type I-525**				
Compressive Strength of Cement (kg/cm2): **550**				
Other Binders Type: -----				
Percent of Usage for Other Binder (%): -----				
w/b: **0.46**		Free Water (L): **191**		
Structural Element: **Foundation, Deck, Column, Wall**				
Target Slump: **190 mm**				
Type of Super-Plasticizer: **Polycarboxylate ether BPC-40**				
Dosage of Super-Plasticizer (%): **0.7**		Water Reducing Rate (%): **21**		

Final Mix

Constituent Material	Weight for 1m3 (SSD)	Weight for 1m3 (Dry)	Weight for lab trial 30L (Dry)	Weight for one batch 1.8m3 (SSD)
Aggregate 11-19	424 kg	420 kg	12.60 kg	763 kg
Aggregate 5-12	202 kg	200 kg	6.00 kg	364 kg
Sand 0-8	588 kg	578 kg	17.34 kg	1058 kg
Sand 0-5	591 kg	582 kg	17.46 kg	1064 kg
Cement	415 kg	415 kg	12.45 kg	747 kg
Other Binder	-----	-----	-----	-----
Water	191 kg	216 kg	6.48 kg	344 kg
Super-plasticizer 0.7%	2.91 kg	2.91 kg	87.3 gr	5.24 kg
Other	-----	-----	-----	-----
Total	2414 kg	2414 kg	72.42 kg	4345 kg
Other Descriptions:				

Table 5.126: **The mix design for C40.2 concrete with company (A) cement**

Batching Plant Name: **Construction Company**

Structural Compressive Strength (MPa): **40**

Standard Deviation (MPa): **2.5**

Mix design Compressive Strength (MPa): **45**

Aggregate Type	Producer	Percentageof Usage (%)	Water Absorption (%)	Water Content (%)
5-12 mm	-----	26	0.98	-----
Sand 0-8	-----	37	1.69	-----
Sand 0-5	-----	37	1.54	-----

Fineness Module of Total Aggregates: **4.19**

Type and Producer of Cement: **Company (A) ASTM Type II**

Compressive Strength of Cement (kg/cm2): **470**

Other Binders Type: -----

Percent of Usage for Other Binder (%): -----

w/b: **0.43**	Free Water (L): **196**

Structural Element: **Deck, Column, Wall**

Target Slump: **190 mm**

Type of Super-Plasticizer: **Polycarboxylate ether BPC-40**

Dosage of Super-Plasticizer (%): **0.7**	Water Reducing Rate (%): **23**

Final Mix

Constituent Material	Weight for 1m3 (SSD)	Weight for 1m3 (Dry)	Weight for lab trial 30L (Dry)	Weight for one batch 0.8m3 (SSD)
Aggregate 5-12	461 kg	456 kg	13.68 kg	369 kg
Sand 0-8	635 kg	624 kg	18.72 kg	508 kg
Sand 0-5	639 kg	629 kg	18.87 kg	511 kg
Cement	465 kg	465 kg	13.95 kg	372 kg
Other Binder	-----	-----	-----	-----
Water	196 kg	221 kg	6.63 kg	157 kg
Super-plasticizer 0.7%	3.19 kg	3.19 kg	95.7 gr	2.55 kg
Other	-----	-----	-----	-----
Total	2399 kg	2398 kg	71.95 kg	1919 kg

Other Descriptions:

Table 5.127: **The mix design for C40.2 concrete with company (B) cement**

Batching Plant Name: **Construction Company**

Structural Compressive Strength (MPa): **40**

Standard Deviation (MPa): **2.5**

Mix design Compressive Strength (MPa): **45**

Aggregate Type	Producer	Percentage of Usage (%)	Water Absorption (%)	Water Content (%)
5-12 mm	-----	26	0.98	-----
Sand 0-8	-----	37	1.69	-----
Sand 0-5	-----	37	1.54	-----

Fineness Module of Total Aggregates: **4.19**

Type and Producer of Cement: **Company(B) ASTM Type I-525**

Compressive Strength of Cement (kg/cm2): **550**

Other Binders Type: -----

Percent of Usage for Other Binder (%): -----

w/b: **0.46**	Free Water (L): **201**

Structural Element: **Deck, Column, Wall**

Target Slump: **190 mm**

Type of Super-Plasticizer: **Polycarboxylate ether BPC-40**

Dosage of Super-Plasticizer (%): **0.7**	Water Reducing Rate (%): **21**

<div align="center">**Final Mix**</div>				
Constituent Material	**Weight for 1m3 (SSD)**	**Weight for 1m3 (Dry)**	**Weight for lab trial 30L (Dry)**	**Weight for one batch 1.8m3 (SSD)**
Aggregate 5-12	463 kg	458 kg	13.74 kg	833 kg
Sand 0-8	639 kg	628 kg	18.84 kg	1150 kg
Sand 0-5	643 kg	633 kg	18.99 kg	1157 kg
Cement	437 kg	437 kg	13.11 kg	787 kg
Other Binder	-----	-----	-----	-----
Water	201 kg	226 kg	6.78 kg	362 kg
Super-plasticizer 0.7%	3.06 kg	3.06 kg	91.8 gr	5.51 kg
Other	-----	-----	-----	-----
Total	2386 kg	2385 kg	71.55 kg	4294 kg

Other Descriptions:

Table 5.128: **The mix design for C40.3 concrete with company (A) cement**				
Batching Plant Name: **Construction Company**				
Structural Compressive Strength (MPa): **40**				
Standard Deviation (MPa): **2.5**				
Mix design Compressive Strength (MPa): **45**				
Aggregate Type	Producer	Percentage of Usage (%)	Water Absorption (%)	Water Content (%)
11-19 mm	-----	23	0.98	-----
5-12 mm	-----	11	0.98	-----
Sand 0-8	-----	33	1.69	-----
Sand 0-5	-----	33	1.54	-----
Fineness Module of Total Aggregates: **4.54**				
Type and Producer of Cement: **Company (A) ASTM Type II**				
Compressive Strength of Cement (kg/cm2): **470**				
Other Binders Type: **GGBS**				
Percent of Usage for Other Binder (%): **20**				
w/b: **0.43**		Free Water (L): **186**		
Structural Element: **Foundation, Deck, Column, Wall**				
Target Slump: **190 mm**				
Type of Super-Plasticizer: **Polycarboxylate ether BPC-40**				
Dosage of Super-Plasticizer (%): **0.7**		Water Reducing Rate (%): **23**		

Final Mix

Constituent Material	Weight for 1m3 (SSD)	Weight for 1m3 (Dry)	Weight for lab trial 30L (Dry)	Weight for one batch 0.8m3 (SSD)
Aggregate 11-19	408 kg	404 kg	12.12 kg	326 kg
Aggregate 5-12	195 kg	193 kg	5.79 kg	156 kg
Sand 0-8	566 kg	556 kg	16.68 kg	453 kg
Sand 0-5	569 kg	560 kg	16.80 kg	455 kg
Cement	398 kg	398 kg	11.94 kg	318 kg
Other Binder (GGBS)	100 kg	100 kg	3.00 kg	80 kg
Water	186 kg	210 kg	6.30 kg	149 kg
Super-plasticizer 0.7%	3.49 kg	3.49 kg	104.7 gr	2.79 kg
Other	-----	-----	-----	-----
Total	2425 kg	2424 kg	72.73 kg	1940 kg
Other Descriptions:				

Table 5.129: **The mix design for C40.3 concrete with company (B) cement**				
Batching Plant Name: **Construction Company**				
Structural Compressive Strength (MPa): **40**				
Standard Deviation (MPa): **2.5**				
Mix design Compressive Strength (MPa): **45**				
Aggregate Type	Producer	Percentage of Usage (%)	Water Absorption (%)	Water Content (%)
11-19 mm	-----	23	0.98	-----
5-12 mm	-----	11	0.98	-----
Sand 0-8	-----	33	1.69	-----
Sand 0-5	-----	33	1.54	-----
Fineness Module of Total Aggregates: **4.54**				
Type and Producer of Cement: **Company (B) ASTM Type I-525**				
Compressive Strength of Cement (kg/cm2): **550**				
Other Binders Type: **GGBS**				
Percent of Usage for Other Binder (%): **20**				
w/b: **0.46**		Free Water (L): **191**		
Structural Element: **Foundation, Deck, Column, Wall**				
Target Slump: **190 mm**				
Type of Super-Plasticizer: **Polycarboxylate ether BPC-40**				
Dosage of Super-Plasticizer (%): **0.7**		Water Reducing Rate (%): **21**		

Final Mix

Constituent Material	Weight for 1m3 (SSD)	Weight for 1m3 (Dry)	Weight for lab trial 30L (Dry)	Weight for one batch 1.8m3 (SSD)
Aggregate 11-19	409 kg	405 kg	12.15 kg	736 kg
Aggregate 5-12	195 kg	193 kg	5.79 kg	351 kg
Sand 0-8	567 kg	557 kg	16.71 kg	1021 kg
Sand 0-5	571 kg	562 kg	16.86 kg	1028 kg
Cement	382 kg	382 kg	11.46 kg	688 kg
Other Binder (GGBS)	95 kg	95 kg	2.85 kg	171 kg
Water	191 kg	215 kg	6.45 kg	344 kg
Super-plasticizer 0.7%	3.34 kg	3.34 kg	100.2 gr	6.01 kg
Other	-----	-----	-----	-----
Total	2413 kg	2412 kg	72.37 kg	4345 kg
Other Descriptions:				

Table 5.130: **The mix design for C40.4 concrete with company (A) cement**				
Batching Plant Name: **Construction Company**				
Structural Compressive Strength (MPa): **40**				
Standard Deviation (MPa): **2.5**				
Mix design Compressive Strength (MPa): **45**				
Aggregate Type	Producer	Percentage of Usage (%)	Water Absorption (%)	Water Content (%)
5-12 mm	-----	26	0.98	-----
Sand 0-8	-----	37	1.69	-----
Sand 0-5	-----	37	1.54	-----
Fineness Module of Total Aggregates: **4.19**				
Type and Producer of Cement: **Company (A) ASTM Type II**				
Compressive Strength of Cement (kg/cm2): **470**				
Other Binders Type: **GGBS**				
Percent of Usage for Other Binder (%): **20**				
w/b: **0.43**		Free Water (L): **196**		
Structural Element: **Deck, Column, Wall**				
Target Slump: **190 mm**				
Type of Super-Plasticizer: **Polycarboxylate ether BPC-40**				
Dosage of Super-Plasticizer (%): **0.7**		Water Reducing Rate (%): **23**		

Final Mix

Constituent Material	Weight for 1m3 (SSD)	Weight for 1m3 (Dry)	Weight for lab trial 30L (Dry)	Weight for one batch 0.8m3 (SSD)
Aggregate 5-12	445 kg	441 kg	13.23 kg	356 kg
Sand 0-8	613 kg	603 kg	18.09 kg	490 kg
Sand 0-5	617 kg	607 kg	18.21 kg	494 kg
Cement	419 kg	419 kg	12.57 kg	335 kg
Other Binder (GGBS)	105 kg	105 kg	3.15 kg	84 kg
Water	196 kg	220 kg	6.60 kg	157 kg
Super-plasticizer 0.7%	3.67 kg	3.67 kg	110.1 gr	2.94 kg
Other	-----	-----	-----	-----
Total	2398 kg	2398 kg	71.96 kg	1919 kg
Other Descriptions:				

Table 5.131: **The mix design for C40.4 concrete with company (B) cement**				
Batching Plant Name: **Construction Company**				
Structural Compressive Strength (MPa): **40**				
Standard Deviation (MPa): **2.5**				
Mix design Compressive Strength (MPa): **45**				
Aggregate Type	Producer	Percentage of Usage (%)	Water Absorption (%)	Water Content (%)
5-12 mm	-----	26	0.98	-----
Sand 0-8	-----	37	1.69	-----
Sand 0-5	-----	37	1.54	-----
Fineness Module of Total Aggregates: **4.19**				
Type and Producer of Cement: **Company (B) ASTM Type I-525**				
Compressive Strength of Cement (kg/cm2): **550**				
Other Binders Type: **GGBS**				
Percent of Usage for Other Binder (%): **20**				
w/b: **0.46**	Free Water (L): **201**			
Structural Element: **Deck, Column, Wall**				
Target Slump: **190 mm**				
Type of Super-Plasticizer: **Polycarboxylate ether BPC-40**				
Dosage of Super-Plasticizer (%): **0.7**	Water Reducing Rate (%): **21**			

Final Mix

Constituent Material	Weight for 1m3 (SSD)	Weight for 1m3 (Dry)	Weight for lab trial 30L (Dry)	Weight for one batch 1.8m3 (SSD)
Aggregate 5-12	446 kg	442 kg	13.26 kg	803 kg
Sand 0-8	615 kg	605 kg	18.15 kg	1107 kg
Sand 0-5	619 kg	609 kg	18.27 kg	1114 kg
Cement	402 kg	402 kg	12.06 kg	724 kg
Other Binder (GGBS)	100 kg	100 kg	3.00 kg	180 kg
Water	201 kg	225 kg	6.75 kg	362 kg
Super-plasticizer 0.7%	3.51 kg	3.51 kg	105.3 gr	6.32 kg
Other	-----	-----	-----	-----
Total	2386 kg	2386 kg	71.59 kg	4296 kg
Other Descriptions:				

Materials Research Forum LLC
https://doi.org/10.21741/9781644900598

Table 5.132: **The mix design for C45.1 concrete with company (A) cement**				
Batching Plant Name: **Construction Company**				
Structural Compressive Strength (MPa): **45**				
Standard Deviation (MPa): **2.5**				
Mix design Compressive Strength (MPa): **50**				
Aggregate Type	Producer	Percentage of Usage (%)	Water Absorption (%)	Water Content (%)
5-12 mm	-----	26	0.98	-----
Sand 0-8	-----	37	1.69	-----
Sand 0-5	-----	37	1.54	-----
Fineness Module of Total Aggregates: **4.19**				
Type and Producer of Cement: **Company (A) ASTM Type II**				
Compressive Strength of Cement (kg/cm2): **470**				
Other Binders Type: -----				
Percent of Usage for Other Binder (%): -----				
w/b: **0.40**		Free Water (L): **193**		
Structural Element: **Deck, Column, Wall**				
Target Slump: **190 mm**				
Type of Super-Plasticizer: **Polycarboxylate ether BPC-40**				
Dosage of Super-Plasticizer (%): **0.8**		Water Reducing Rate (%): **25**		

Final Mix

Constituent Material	Weight for 1m3 (SSD)	Weight for 1m3 (Dry)	Weight for lab trial 30L (Dry)	Weight for one batch 0.8m3 (SSD)
Aggregate 5-12	458 kg	454 kg	13.62 kg	366 kg
Sand 0-8	632 kg	621 kg	18.63 kg	506 kg
Sand 0-5	636 kg	626 kg	18.78 kg	509 kg
Cement	482 kg	482 kg	14.46 kg	386 kg
Other Binder	-----	-----	-----	-----
Water	193 kg	218 kg	6.54 kg	154 kg
Super-plasticizer 0.8%	3.86 kg	3.86 kg	115.8 gr	3.09 kg
Other	-----	-----	-----	-----
Total	2405 kg	2405 kg	72.14 kg	1924 kg
Other Descriptions:				

Table 5.133: **The mix design for C45.1 concrete with company (B) cement**				
Batching Plant Name: **Construction Company**				
Structural Compressive Strength (MPa): **45**				
Standard Deviation (MPa): **2.5**				
Mix design Compressive Strength (MPa): **50**				
Aggregate Type	Producer	Percentage of Usage (%)	Water Absorption (%)	Water Content (%)
5-12 mm	-----	26	0.98	-----
Sand 0-8	-----	37	1.69	-----
Sand 0-5	-----	37	1.54	-----
Fineness Module of Total Aggregates: **4.19**				
Type and Producer of Cement: **Company (B) ASTM Type I-525**				
Compressive Strength of Cement (kg/cm2): **550**				
Other Binders Type: -----				
Percent of Usage for Other Binder (%): -----				
w/b: **0.44**		Free Water (L): **199**		
Structural Element: **Deck, Column, Wall**				
Target Slump: **190 mm**				
Type of Super-Plasticizer: **Polycarboxylate ether BPC-40**				
Dosage of Super-Plasticizer (%): **0.8**		Water Reducing Rate (%): **23**		
Final Mix				
Constituent Material	**Weight for 1m3 (SSD)**	**Weight for 1m3 (Dry)**	**Weight for lab trial 30L (Dry)**	**Weight for one batch 1.8m3 (SSD)**
Aggregate 5-12	461 kg	456 kg	13.68 kg	830 kg
Sand 0-8	635 kg	624 kg	18.72 kg	1143 kg
Sand 0-5	639 kg	629 kg	18.87 kg	1150 kg
Cement	453 kg	453 kg	13.59 kg	815 kg
Other Binder	-----	-----	-----	-----
Water	199 kg	224 kg	6.72 kg	358 kg
Super-plasticizer 0.8%	3.62 kg	3.62 kg	108.6 gr	6.52 kg
Other	-----	-----	-----	-----
Total	2390 kg	2389 kg	71.68 kg	4302 kg
Other Descriptions:				

Table 5.134: **The mix design for C45.2 concrete with company (A) cement**				
Batching Plant Name: **Construction Company**				
Structural Compressive Strength (MPa): **45**				
Standard Deviation (MPa): **2.5**				
Mix design Compressive Strength (MPa): **50**				
Aggregate Type	Producer	Percentage of Usage (%)	Water Absorption (%)	Water Content (%)
5-12 mm	-----	26	0.98	-----
Sand 0-8	-----	37	1.69	-----
Sand 0-5	-----	37	1.54	-----
Fineness Module of Total Aggregates: **4.19**				
Type and Producer of Cement: **Company (A) ASTM Type II**				
Compressive Strength of Cement (kg/cm2): **470**				
Other Binders Type: **GGBS**				
Percent of Usage for Other Binder (%): **20**				
w/b: **0.40**		Free Water (L): **193**		
Structural Element: **Deck, Column, Wall**				
Target Slump: **190 mm**				
Type of Super-Plasticizer: **Polycarboxylate ether BPC-40**				
Dosage of Super-Plasticizer (%): **0.8**		Water Reducing Rate (%): **25**		

Final Mix				
Constituent Material	**Weight for 1m3 (SSD)**	**Weight for 1m3 (Dry)**	**Weight for lab trial 30L (Dry)**	**Weight for one batch 0.8m3 (SSD)**
Aggregate 5-12	**439 kg**	**435 kg**	**13.05 kg**	**351 kg**
Sand 0-8	**605 kg**	**595 kg**	**17.85 kg**	**484 kg**
Sand 0-5	**609 kg**	**600 kg**	**18.00 kg**	**487 kg**
Cement	**443 kg**	**443 kg**	**13.29 kg**	**354 kg**
Other Binder (GGBS)	**111 kg**	**111 kg**	**3.33 kg**	**89 kg**
Water	**193 kg**	**217 kg**	**6.51 kg**	**154 kg**
Super-plasticizer 0.8%	**4.43 kg**	**4.43 kg**	**132.9 gr**	**3.54 kg**
Other	-----	-----	-----	-----
Total	**2404 kg**	**2405 kg**	**72.16 kg**	**1922 kg**
Other Descriptions:				

Table 5.135: **The mix design for C45.2 concrete with company (B) cement**				
Batching Plant Name: **Construction Company**				
Structural Compressive Strength (MPa): **45**				
Standard Deviation (MPa): **2.5**				
Mix design Compressive Strength (MPa): **50**				
Aggregate Type	Producer	Percentage of Usage (%)	Water Absorption (%)	Water Content (%)
5-12 mm	-----	26	0.98	-----
Sand 0-8	-----	37	1.69	-----
Sand 0-5	-----	37	1.54	-----
Fineness Module of Total Aggregates: **4.19**				
Type and Producer of Cement: **Company (B) ASTM Type I-525**				
Compressive Strength of Cement (kg/cm2): **550**				
Other Binders Type: **GGBS**				
Percent of Usage for Other Binder (%): **20**				
w/b: **0.44**		Free Water (L): **199**		
Structural Element: **Deck, Column, Wall**				
Target Slump: **190 mm**				
Type of Super-Plasticizer: **Polycarboxylate ether BPC-40**				
Dosage of Super-Plasticizer (%): **0.8**		Water Reducing Rate (%): **23**		

Final Mix				
Constituent Material	**Weight for 1m3 (SSD)**	**Weight for 1m3 (Dry)**	**Weight for lab trial 30L (Dry)**	**Weight for one batch 1.8m3 (SSD)**
Aggregate 5-12	442 kg	438 kg	13.14 kg	796 kg
Sand 0-8	610 kg	600 kg	18.00 kg	1098 kg
Sand 0-5	614 kg	605 kg	18.15 kg	1105 kg
Cement	417 kg	417 kg	12.51 kg	751 kg
Other Binder (GGBS)	104 kg	104 kg	3.12 kg	187 kg
Water	199 kg	223 kg	6.69 kg	358 kg
Super-plasticizer 0.8%	4.17 kg	4.17 kg	125.1 gr	7.51 kg
Other	-----	-----	-----	-----
Total	2390 kg	2391 kg	71.73 kg	4302 kg
Other Descriptions:				

Table 5.136: **The mix design for C50.1 concrete with company (A) cement**				
Batching Plant Name: **Construction Company**				
Structural Compressive Strength (MPa): **50**				
Standard Deviation (MPa): **2.5**				
Mix design Compressive Strength (MPa): **55**				
Aggregate Type	Producer	Percentage of Usage (%)	Water Absorption (%)	Water Content (%)
5-12 mm	-----	26	0.98	-----
Sand 0-8	-----	37	1.69	-----
Sand 0-5	-----	37	1.54	-----
Fineness Module of Total Aggregates: **4.19**				
Type and Producer of Cement: **Company (A) ASTM Type II**				
Compressive Strength of Cement (kg/cm2): **470**				
Other Binders Type: **GGBS**				
Percent of Usage for Other Binder (%): **20**				
w/b: **0.37**		Free Water (L): **191**		
Structural Element: **Deck, Column, Wall**				
Target Slump: **190 mm**				
Type of Super-Plasticizer: **Polycarboxylate ether BPC-40**				
Dosage of Super-Plasticizer (%): **0.9**		Water Reducing Rate (%): **27**		

Final Mix

Constituent Material	Weight for 1m3 (SSD)	Weight for 1m3 (Dry)	Weight for lab trial 30L (Dry)	Weight for one batch 0.8m3 (SSD)
Aggregate 5-12	431 kg	427 kg	12.81 kg	345 kg
Sand 0-8	594 kg	584 kg	17.52 kg	475 kg
Sand 0-5	597 kg	588 kg	17.64 kg	478 kg
Cement	474 kg	474 kg	14.22 kg	379 kg
Other Binder (GGBS)	119 kg	119 kg	3.57 kg	95 kg
Water	191 kg	214 kg	6.42 kg	153 kg
Super-plasticizer 0.9%	5.34 kg	5.34 kg	160.2 gr	4.27 kg
Other	-----	-----	-----	-----
Total	2411 kg	2411 kg	72.34 kg	1929 kg
Other Descriptions:				

Table 5.137: **The mix design for C50.1 concrete with company (B) cement**				
Batching Plant Name: **Construction Company**				
Structural Compressive Strength (MPa): **50**				
Standard Deviation (MPa): **2.5**				
Mix design Compressive Strength (MPa): **55**				
Aggregate Type	Producer	Percentage of Usage (%)	Water Absorption (%)	Water Content (%)
5-12 mm	-----	26	0.98	-----
Sand 0-8	-----	37	1.69	-----
Sand 0-5	-----	37	1.54	-----
Fineness Module of Total Aggregates: **4.19**				
Type and Producer of Cement: **Company (B) ASTM Type I-525**				
Compressive Strength of Cement (kg/cm2): **550**				
Other Binders Type: **GGBS**				
Percent of Usage for Other Binder (%): **20**				
w/b: **0.41**		Free Water (L): **196**		
Structural Element: **Deck, Column, Wall**				
Target Slump: **190 mm**				
Type of Super-Plasticizer: **Polycarboxylate ether BPC-40**				
Dosage of Super-Plasticizer (%): **0.9**		Water Reducing Rate (%): **25**		

Final Mix				
Constituent Material	**Weight for 1m3 (SSD)**	**Weight for 1m3 (Dry)**	**Weight for lab trial 30L (Dry)**	**Weight for one batch 1.8m3 (SSD)**
Aggregate 5-12	437 kg	433 kg	12.99 kg	787 kg
Sand 0-8	603 kg	593 kg	17.79 kg	1085 kg
Sand 0-5	607 kg	598 kg	17.94 kg	1093 kg
Cement	440 kg	440 kg	13.20 kg	792 kg
Other Binder (GGBS)	110 kg	110 kg	3.30 kg	198 kg
Water	196 kg	220 kg	6.60 kg	353 kg
Super-plasticizer 0.9%	4.95 kg	4.95 kg	148.5 gr	8.91 kg
Other	-----	-----	-----	-----
Total	2398 kg	2399 kg	71.97 kg	4317 kg
Other Descriptions:				

Table 5.138: **The mix design for C50.2 concrete with company (A) cement**				
Batching Plant Name: **Construction Company**				
Structural Compressive Strength (MPa): **50**				
Standard Deviation (MPa): **2.5**				
Mix design Compressive Strength (MPa): **55**				
Aggregate Type	Producer	Percentage of Usage (%)	Water Absorption (%)	Water Content (%)
5-12 mm	-----	26	0.98	-----
Sand 0-8	-----	37	1.69	-----
Sand 0-5	-----	37	1.54	-----
Fineness Module of Total Aggregates: **4.19**				
Type and Producer of Cement: **Company (A) ASTM Type II**				
Compressive Strength of Cement (kg/cm2): **470**				
Other Binders Type: **GGBS, Silica Fume**				
Percent of Usage for Other Binder (%): **20, 5**				
w/b: **0.37**		Free Water (L): **191**		
Structural Element: **Deck, Column, Wall**				
Target Slump: **190 mm**				
Type of Super-Plasticizer: **Polycarboxylate ether BPC-40**				
Dosage of Super-Plasticizer (%): **0.9**		Water Reducing Rate (%): **27**		
Final Mix				
Constituent Material	**Weight for 1m3 (SSD)**	**Weight for 1m3 (Dry)**	**Weight for lab trial 30L (Dry)**	**Weight for one batch 0.8m3 (SSD)**
Aggregate 5-12	**429 kg**	**425 kg**	**12.75 kg**	**343 kg**
Sand 0-8	**591 kg**	**581 kg**	**17.43 kg**	**473 kg**
Sand 0-5	**594 kg**	**585 kg**	**17.55 kg**	**475 kg**
Cement	**444 kg**	**444 kg**	**13.32 kg**	**355 kg**
Other Binder (GGBS)	**119 kg**	**119 kg**	**3.57 kg**	**95 kg**
Other Binder (Silica Fume)	**30 kg**	**30 kg**	**0.9 kg**	**24 kg**
Water	**191 kg**	**214 kg**	**6.42 kg**	**153 kg**
Super-plasticizer 0.9%	**5.34 kg**	**5.34 kg**	**160.2 gr**	**4.27 kg**
Other	-----	-----	-----	-----
Total	**2403 kg**	**2403 kg**	**72.10 kg**	**1922 kg**
Other Descriptions:				

Table 5.139: **The mix design for C50.2 concrete with company (B) cement**

Batching Plant Name: **Construction Company**				
Structural Compressive Strength (MPa): **50**				
Standard Deviation (MPa): **2.5**				
Mix design Compressive Strength (MPa): **55**				
Aggregate Type	Producer	Percentage of Usage (%)	Water Absorption (%)	Water Content (%)
5-12 mm	-----	26	0.98	-----
Sand 0-8	-----	37	1.69	-----
Sand 0-5	-----	37	1.54	-----
Fineness Module of Total Aggregates: **4.19**				
Type and Producer of Cement: **Company (B) ASTM Type I-525**				
Compressive Strength of Cement (kg/cm2): **550**				
Other Binders Type: **GGBS, Silica Fume**				
Percent of Usage for Other Binder (%): **20, 5**				
w/b: **0.41**		Free Water (L): **196**		
Structural Element: **Deck, Column, Wall**				
Target Slump: **190 mm**				
Type of Super-Plasticizer: **Polycarboxylate ether BPC-40**				
Dosage of Super-Plasticizer (%): **0.9**		Water Reducing Rate (%): **25**		

Final Mix				
Constituent Material	**Weight for 1m3 (SSD)**	**Weight for 1m3 (Dry)**	**Weight for lab trial 30L (Dry)**	**Weight for one batch 1.8m3 (SSD)**
Aggregate 5-12	435 kg	431 kg	12.93 kg	783 kg
Sand 0-8	600 kg	590 kg	17.70 kg	1080 kg
Sand 0-5	604 kg	595 kg	17.85 kg	1087 kg
Cement	413 kg	413 kg	12.39 kg	743 kg
Other Binder (GGBS)	110 kg	110 kg	3.30 kg	198 kg
Other Binder (Silica Fume)	27 kg	27 kg	0.81 kg	49 kg
Water	196 kg	220 kg	6.60 kg	353 kg
Super-plasticizer 0.9%	4.95 kg	4.95 kg	148.5 gr	8.91 kg
Other	-----	-----	-----	-----
Total	2390 kg	2391 kg	71.73 kg	4302 kg
Other Descriptions:				

Table 5.140: **The mix design for C60 concrete with company (A) cement**

Batching Plant Name: **Construction Company**

Structural Compressive Strength (MPa): **60**

Standard Deviation (MPa): **2.5**

Mix design Compressive Strength (MPa): **65**

Aggregate Type	Producer	Percent of Usage (%)	Water Absorption (%)	Water Content (%)
5-12 mm	-----	26	0.98	-----
Sand 0-8	-----	37	1.69	-----
Sand 0-5	-----	37	1.54	-----

Fineness Module of Total Aggregates: **4.19**

Type and Producer of Cement: **Company (A) ASTM Type II**

Compressive Strength of Cement (kg/cm2): **470**

Other Binders Type: **GGBS, Silica Fume**

Percent of Usage for Other Binder (%): **20, 7**

w/b: **0.34** | Free Water (L): **186**

Structural Element: **Deck, Column, Wall**

Target Slump: **190 mm**

Type of Super-Plasticizer: **Polycarboxylate ether BPC-40**

Dosage of Super-Plasticizer (%): **1.1** | Water Reducing Rate (%): **30**

Final Mix

Constituent Material	Weight for 1m3 (SSD)	Weight for 1m3 (Dry)	Weight for lab trial 30L (Dry)	Weight for one batch 0.8m3 (SSD)
Aggregate 5-12	421 kg	417 kg	12.51 kg	337 kg
Sand 0-8	581 kg	571 kg	17.13 kg	465 kg
Sand 0-5	584 kg	575 kg	17.25 kg	467 kg
Cement	459 kg	459 kg	13.77 kg	367 kg
Other Binder (GGBS)	126 kg	126 kg	3.78 kg	101 kg
Other Binder (Silica Fume)	44 kg	44 kg	1.32 kg	35 kg
Water	186 kg	209 kg	6.27 kg	149 kg
Super-plasticizer 1.1%	6.92 kg	6.92 kg	207.6 gr	5.54 kg
Other	-----	-----	-----	-----
Total	2408 kg	2408 kg	72.23 kg	1926 kg

Other Descriptions:

Table 5.141: **The mix design for C60 concrete with company (B) cement**				
Batching Plant Name: **Construction Company**				
Structural Compressive Strength (MPa): **60**				
Standard Deviation (MPa): **2.5**				
Mix design Compressive Strength (MPa): **65**				
Aggregate Type	Producer	Percentage of Usage (%)	Water Absorption (%)	Water Content (%)
5-12 mm	-----	26	0.98	-----
Sand 0-8	-----	37	1.69	-----
Sand 0-5	-----	37	1.54	-----
Fineness Module of Total Aggregates: **4.19**				
Type and Producer of Cement: **Company (B) ASTM Type I-525**				
Compressive Strength of Cement (kg/cm2): **550**				
Other Binders Type: **GGBS, Silica Fume**				
Percent of Usage for Other Binder (%): **20, 7**				
w/b: **0.37**		Free Water (L): **191**		
Structural Element: **Deck, Column, Wall**				
Target Slump: **190 mm**				
Type of Super-Plasticizer: **Polycarboxylate ether BPC-40**				
Dosage of Super-Plasticizer (%): **1.1**		Water Reducing Rate (%): **28**		

Final Mix				
Constituent Material	**Weight for 1m3 (SSD)**	**Weight for 1m3 (Dry)**	**Weight for lab trial 30L (Dry)**	**Weight for one batch 1.8m3 (SSD)**
Aggregate 5-12	426 kg	422 kg	12.66 kg	767 kg
Sand 0-8	588 kg	578 kg	17.34 kg	1058 kg
Sand 0-5	591 kg	582 kg	17.46 kg	1064 kg
Cement	432 kg	432 kg	12.96 kg	778 kg
Other Binder (GGBS)	119 kg	119 kg	3.57 kg	214 kg
Other Binder (Silica Fume)	42 kg	42 kg	1.26 kg	76 kg
Water	191 kg	214 kg	6.42 kg	344 kg
Super-plasticizer 1.1%	6.52 kg	6.52 kg	195.6 gr	11.74 kg
Other	-----	-----	-----	-----
Total	2395 kg	2395 kg	71.86 kg	4313 kg
Other Descriptions:				

Materials Research Forum LLC
https://doi.org/10.21741/9781644900598

Table 5.142: **The mix design for C70 concrete with company (A) cement**				
Batching Plant Name: **Construction Company**				
Structural Compressive Strength (MPa): **70**				
Standard Deviation (MPa): **2.5**				
Mix design Compressive Strength (MPa): **75**				
Aggregate Type	Producer	Percentage of Usage (%)	Water Absorption (%)	Water Content (%)
5-12 mm	-----	26	0.98	-----
Sand 0-8	-----	37	1.69	-----
Sand 0-5	-----	37	1.54	-----
Fineness Module of Total Aggregates: **4.19**				
Type and Producer of Cement: **Company (A) ASTM Type II**				
Compressive Strength of Cement (kg/cm2): **470**				
Other Binders Type: **GGBS, Silica Fume**				
Percent of Usage for Other Binder (%): **20, 9**				
w/b: **0.31**		Free Water (L): **181**		
Structural Element: **Deck, Column, Wall**				
Target Slump: **190 mm**				
Type of Super-Plasticizer: **Polycarboxylate ether BPC-40**				
Dosage of Super-Plasticizer (%): **1.3**		Water Reducing Rate (%): **33**		
Final Mix				
Constituent Material	**Weight for 1m3 (SSD)**	**Weight for 1m3 (Dry)**	**Weight for lab trial 30L (Dry)**	**Weight for one batch 0.8m3 (SSD)**
Aggregate 5-12	**412 kg**	**408 kg**	**12.24 kg**	**330 kg**
Sand 0-8	**568 kg**	**558 kg**	**16.74 kg**	**454 kg**
Sand 0-5	**571 kg**	**562 kg**	**16.86 kg**	**457 kg**
Cement	**477 kg**	**477 kg**	**14.31 kg**	**382 kg**
Other Binder (GGBS)	**134 kg**	**134 kg**	**4.02 kg**	**107 kg**
Other Binder (Silica Fume)	**61 kg**	**61 kg**	**1.83 kg**	**49 kg**
Water	**181 kg**	**203 kg**	**6.09 kg**	**145 kg**
Super-plasticizer 1.3%	**8.74 kg**	**8.74 kg**	**262.2 gr**	**6.99 kg**
Other	-----	-----	-----	-----
Total	**2413 kg**	**2412 kg**	**72.35 kg**	**1931 kg**
Other Descriptions:				

Table 5.143: **The mix design for C70 concrete with company (B) cement**

Batching Plant Name: **Construction Company**

Structural Compressive Strength (MPa): **70**

Standard Deviation (MPa): **2.5**

Mix design Compressive Strength (MPa): **75**

Aggregate Type	Producer	Percentage of Usage (%)	Water Absorption (%)	Water Content (%)
5-12 mm	-----	26	0.98	-----
Sand 0-8	-----	37	1.69	-----
Sand 0-5	-----	37	1.54	-----

Fineness Module of Total Aggregates: **4.19**

Type and Producer of Cement: **Company (B) ASTM Type I-525**

Compressive Strength of Cement (kg/cm2): **550**

Other Binders Type: **GGBS, Silica Fume**

Percent of Usage for Other Binder (%): **20, 9**

w/b: **0.34** | Free Water (L): **186**

Structural Element: **Deck, Column, Wall**

Target Slump: **190 mm**

Type of Super-Plasticizer: **Polycarboxylate ether BPC-40**

Dosage of Super-Plasticizer (%): **1.3** | Water Reducing Rate (%): **31**

Final Mix

Constituent Material	Weight for 1m3 (SSD)	Weight for 1m3 (Dry)	Weight for lab trial 30L (Dry)	Weight for one batch 1.8m3 (SSD)
Aggregate 5-12	418 kg	414 kg	12.42 kg	752 kg
Sand 0-8	577 kg	567 kg	17.01 kg	1039 kg
Sand 0-5	581 kg	572 kg	17.16 kg	1046 kg
Cement	446 kg	446 kg	13.38 kg	803 kg
Other Binder (GGBS)	126 kg	126 kg	3.78 kg	227 kg
Other Binder (Silica Fume)	57 kg	57 kg	1.71 kg	103 kg
Water	186 kg	209 kg	6.27 kg	335 kg
Super-plasticizer 1.3%	8.18 kg	8.18 kg	245.4 gr	14.72 kg
Other	-----	-----	-----	-----
Total	2399 kg	2399 kg	71.97 kg	4319 kg

Other Descriptions:

6.   Confirmation tests for mix designs

Now it is time to test and control the mix designs to see if they satisfy our demands for each kind of concrete. To do that, first we start with a precise laboratory test for each concrete mix design. Then we use the batching plants to produce each kind of concrete industrially, and control the mix designs during a period of time.

6.1   Laboratory tests for mix design control

For laboratory tests, we used a concrete lab mixer with the quantity of 50 L (Fig. 6.1), but we only used 30 L volume. We also used iron cube and cylinder molds (Fig. 6.2) and a 300 ton digital test machine (Fig. 6.3).

Figure 6.1: Laboratory mixer for making trials

Figure 6.2: Iron cube and cylinder molds

Figure 6.3: Digital 300 ton concrete testing machine

For aggregates and cement, we used the same samples that we are going to use in the batching plant (Fig. 6.4).

Figure 6.4: Sand sample in the batching plant

To do the test more precisely, we first dried all of the aggregates in the oven, let them cool down, and then we used them for laboratory trials. For each concrete, we made one trial. We used a precise amount for each kind of constituent material. Then we checked initial slump, slump retention after 30 minutes, concrete temperature and concrete compressive strength at seven and twenty-eight days and also hardened concrete specific gravity.

6.1.1   *Laboratory trials results*

You can see the results of the laboratory trials in tables 6.1 to 6.40.

Table 6.1: **Laboratory test results for C25.1 made with company (A) cement**		
Initial slump: **185 mm**	Final slump (After 30 minutes): **160 mm**	
Initial concrete temperature: **25.3 °C**	Final concrete temperature: **26.9 °C**	
Concrete Age (days)	Specific Gravity (kg/m3)	Compressive Strength (MPa)
7	2391	25.8
7	2382	26.1
28	2379	31.1
28	2403	32.8
Mean value for specific gravity (kg/m3)	2389	Mean value for compressive strength (MPa) **32.0**

Table 6.2: **Laboratory test results for C25.1 made with company (B) cement**		
Initial slump: **160 mm**	Final slump (After 30 minutes): **120 mm**	
Initial concrete temperature: **24.3 °C**	Final concrete temperature: **25.0 °C**	
Concrete Age (days)	Specific Gravity (kg/m3)	Compressive Strength (MPa)
7	2373	27.6
7	2381	26.9
28	2369	30.1
28	2389	30.9
Mean value for specific gravity (kg/m3)	2378	Mean value for compressive strength (MPa) **30.5**

Table 6.3: **Laboratory test results for C25.2 made with company (A) cement**		
Initial slump: **190 mm**	Final slump (After 30 minutes): **170 mm**	
Initial concrete temperature: **23.8 °C**	Final concrete temperature: **24.5 °C**	
Concrete Age (days)	Specific Gravity (kg/m3)	Compressive Strength (MPa)
7	2380	25.9
7	2383	24.8
28	2378	31.0
28	2385	31.5
Mean value for specific gravity (kg/m3)	2381	Mean value for compressive strength (MPa) **31.2**

Table 6.4: **Laboratory test results for C25.2 made with company (B) cement**

Initial slump: **180 mm**		Final slump (After 30 minutes): **165 mm**	
Initial concrete temperature: **23.5 °C**		Final concrete temperature: **25.1 °C**	
Concrete Age (days)	Specific Gravity (kg/m3)	Compressive Strength (MPa)	
7	2379	26.6	
7	2373	26.1	
28	2381	31.2	
28	2372	30.4	
Mean value for specific gravity (kg/m3)	**2376**	Mean value for compressive strength (MPa)	**30.8**

Table 6.5: **Laboratory test results for C30.1 made with company (A) cement**

Initial slump: **195 mm**		Final slump (After 30 minutes): **185 mm**	
Initial concrete temperature: **24.3 °C**		Final concrete temperature: **26.2 °C**	
Concrete Age (days)	Specific Gravity (kg/m3)	Compressive Strength (MPa)	
7	2391	29.9	
7	2399	30.9	
28	2395	37.1	
28	2390	36.9	
Mean value for specific gravity (kg/m3)	**2394**	Mean value for compressive strength (MPa)	**37.0**

Table 6.6: **Laboratory test results for C30.1 made with company (B) cement**

Initial slump: **185 mm**		Final slump (After 30 minutes): **170 mm**	
Initial concrete temperature: **23.3 °C**		Final concrete temperature: **25.0 °C**	
Concrete Age (days)	Specific Gravity (kg/m3)	Compressive Strength (MPa)	
7	2381	32.1	
7	2377	31.8	
28	2386	35.5	
28	2388	36.1	
Mean value for specific gravity (kg/m3)	**2383**	Mean value for compressive strength (MPa)	**35.8**

Table 6.7: Laboratory test results for C30.2 made with company (A) cement

Initial slump: **195 mm**		Final slump (After 30 minutes): **190 mm**	
Initial concrete temperature: **23.1 °C**		Final concrete temperature: **24.9 °C**	
Concrete Age (days)	Specific Gravity (kg/m3)		Compressive Strength (MPa)
7	2365		30.1
7	2369		29.5
28	2371		35.1
28	2361		35.5
Mean value for specific gravity (kg/m3)	2366	Mean value for compressive strength (MPa)	35.3

Table 6.8: Laboratory test results for C30.2 made with company (B) cement

Initial slump: **190 mm**		Final slump (After 30 minutes): **170 mm**	
Initial concrete temperature: **23.1 °C**		Final concrete temperature: **25.6 °C**	
Concrete Age (days)	Specific Gravity (kg/m3)		Compressive Strength (MPa)
7	2361		29.8
7	2355		31.1
28	2369		35.1
28	2361		35.8
Mean value for specific gravity (kg/m3)	2361	Mean value for compressive strength (MPa)	35.4

Table 6.9: Laboratory test results for C30.3 made with company (A) cement

Initial slump: **180 mm**		Final slump (After 30 minutes): **145 mm**	
Initial concrete temperature: **24.1 °C**		Final concrete temperature: **26.2 °C**	
Concrete Age (days)	Specific Gravity (kg/m3)		Compressive Strength (MPa)
7	2395		23.4
7	2401		22.9
28	2390		35.5
28	2408		36.8
Mean value for specific gravity (kg/m3)	2398	Mean value for compressive strength (MPa)	36.1

Table 6.10: **Laboratory test results for C30.3 made with company (B) cement**

Initial slump: **175 mm**		Final slump (After 30 minutes): **130 mm**	
Initial concrete temperature: **23.1 °C**		Final concrete temperature: **26.2 °C**	
Concrete Age (days)	Specific Gravity (kg/m3)	Compressive Strength (MPa)	
7	2401	21.8	
7	2408	22.3	
28	2399	35.1	
28	2406	36.2	
Mean value for specific gravity (kg/m3)	2403	Mean value for compressive strength (MPa)	35.6

Table 6.11: **Laboratory test results for C30.4 made with company (A) cement**

Initial slump: **190 mm**		Final slump (After 30 minutes): **145 mm**	
Initial concrete temperature: **21.6 °C**		Final concrete temperature: **23.3 °C**	
Concrete Age (days)	Specific Gravity (kg/m3)	Compressive Strength (MPa)	
7	2371	22.9	
7	2370	23.8	
28	2376	35.6	
28	2368	36.1	
Mean value for specific gravity (kg/m3)	2371	Mean value for compressive strength (MPa)	35.8

Table 6.12: **Laboratory test results for C30.4 made with company (B) cement**

Initial slump: **185 mm**		Final slump (After 30 minutes): **130 mm**	
Initial concrete temperature: **22.2 °C**		Final concrete temperature: **24.1 °C**	
Concrete Age (days)	Specific Gravity (kg/m3)	Compressive Strength (MPa)	
7	2368	23.1	
7	2360	23.6	
28	2359	36.1	
28	2371	37.2	
Mean value for specific gravity (kg/m3)	2364	Mean value for compressive strength (MPa)	36.6

Table 6.13: **Laboratory test results for C35.1 made with company (A) cement**

Initial slump: **190 mm**		Final slump (After 30 minutes): **175 mm**	
Initial concrete temperature: **24.1 °C**		Final concrete temperature: **26.0 °C**	
Concrete Age (days)	Specific Gravity (kg/m3)		Compressive Strength (MPa)
7	2412		34.9
7	2405		35.3
28	2409		40.8
28	2401		41.1
Mean value for specific gravity (kg/m3)	2407	Mean value for compressive strength (MPa)	40.9

Table 6.14: **Laboratory test results for C35.1 made with company (B) cement**

Initial slump: **180 mm**		Final slump (After 30 minutes): **165 mm**	
Initial concrete temperature: **23.5 °C**		Final concrete temperature: **25.9 °C**	
Concrete Age (days)	Specific Gravity (kg/m3)		Compressive Strength (MPa)
7	2399		36.1
7	2391		35.5
28	2402		40.9
28	2395		41.1
Mean value for specific gravity (kg/m3)	2397	Mean value for compressive strength (MPa)	41.0

Table 6.15: **Laboratory test results for C35.2 made with company (A) cement**

Initial slump: **205 mm**		Final slump (After 30 minutes): **200 mm**	
Initial concrete temperature: **21.1 °C**		Final concrete temperature: **22.9 °C**	
Concrete Age (days)	Specific Gravity (kg/m3)		Compressive Strength (MPa)
7	2381		34.5
7	2370		35.0
28	2379		40.3
28	2375		39.8
Mean value for specific gravity (kg/m3)	2376	Mean value for compressive strength (MPa)	40.0

Table 6.16: **Laboratory test results for C35.2 made with company (B) cement**			
Initial slump: **200 mm**		Final slump (After 30 minutes): **185 mm**	
Initial concrete temperature: **26.1 °C**		Final concrete temperature: **27.5 °C**	
Concrete Age (days)	Specific Gravity (kg/m3)	Compressive Strength (MPa)	
7	2372	36.9	
7	2370	35.6	
28	2381	40.9	
28	2379	41.7	
Mean value for specific gravity (kg/m3)	2375	Mean value for compressive strength (MPa)	**41.3**

Table 6.17: **Laboratory test results for C35.3 made with company (A) cement**			
Initial slump: **195 mm**		Final slump (After 30 minutes): **145 mm**	
Initial concrete temperature: **22.2 °C**		Final concrete temperature: **24.1 °C**	
Concrete Age (days)	Specific Gravity (kg/m3)	Compressive Strength (MPa)	
7	2401	32.5	
7	2399	33.1	
28	2386	40.5	
28	2410	40.0	
Mean value for specific gravity (kg/m3)	2399	Mean value for compressive strength (MPa)	**40.2**

Table 6.18: **Laboratory test results for C35.3 made with company (B) cement**			
Initial slump: **190 mm**		Final slump (After 30 minutes): **130 mm**	
Initial concrete temperature: **23.4 °C**		Final concrete temperature: **25.2 °C**	
Concrete Age (days)	Specific Gravity (kg/m3)	Compressive Strength (MPa)	
7	2410	33.1	
7	2405	32.8	
28	2408	39.5	
28	2401	40.9	
Mean value for specific gravity (kg/m3)	2406	Mean value for compressive strength (MPa)	**40.2**

Table 6.19: Laboratory test results for C35.4 made with company (A) cement

Initial slump: **205 mm**		Final slump (After 30 minutes): **160 mm**	
Initial concrete temperature: **24.5 °C**		Final concrete temperature: **26.2 °C**	
Concrete Age (days)	Specific Gravity (kg/m3)	Compressive Strength (MPa)	
7	2381	31.8	
7	2379	32.5	
28	2375	39.7	
28	2393	39.9	
Mean value for specific gravity (kg/m3)	**2382**	Mean value for compressive strength (MPa)	**39.8**

Table 6.20: Laboratory test results for C35.4 made with company (B) cement

Initial slump: **190 mm**		Final slump (After 30 minutes): **145 mm**	
Initial concrete temperature: **24.5 °C**		Final concrete temperature: **26.1 °C**	
Concrete Age (days)	Specific Gravity (kg/m3)	Compressive Strength (MPa)	
7	2375	33.1	
7	2379	32.7	
28	2371	40.5	
28	2376	40.8	
Mean value for specific gravity (kg/m3)	**2375**	Mean value for compressive strength (MPa)	**40.6**

Table 6.21: Laboratory test results for C40.1 made with company (A) cement

Initial slump: **205 mm**		Final slump (After 30 minutes): **195 mm**	
Initial concrete temperature: **21.1 °C**		Final concrete temperature: **22 5 °C**	
Concrete Age (days)	Specific Gravity (kg/m3)	Compressive Strength (MPa)	
7	2410	39.5	
7	2421	40.8	
28	2415	46.1	
28	2411	47.1	
Mean value for specific gravity (kg/m3)	**2414**	Mean value for compressive strength (MPa)	**46.6**

Table 6.22: **Laboratory test results for C40.1 made with company (B) cement**

Initial slump: **210 mm**		Final slump (After 30 minutes): **190 mm**	
Initial concrete temperature: **22.1 °C**		Final concrete temperature: **23.8 °C**	
Concrete Age (days)	Specific Gravity (kg/m3)	Compressive Strength (MPa)	
7	2410	40.5	
7	2405	40.8	
28	2401	47.1	
28	2409	46.5	
Mean value for specific gravity (kg/m3)	**2406**	Mean value for compressive strength (MPa)	**46.8**

Table 6.23: **Laboratory test results for C40.2 made with company (A) cement**

Initial slump: **200 mm**		Final slump (After 30 minutes): **200 mm**	
Initial concrete temperature: **20.8 °C**		Final concrete temperature: **22.1 °C**	
Concrete Age (days)	Specific Gravity (kg/m3)	Compressive Strength (MPa)	
7	2385	40.9	
7	2388	39.5	
28	2381	45.8	
28	2385	46.7	
Mean value for specific gravity (kg/m3)	**2385**	Mean value for compressive strength (MPa)	**46.2**

Table 6.24: **Laboratory test results for C40.2 made with company (B) cement**

Initial slump: **205 mm**		Final slump (After 30 minutes): **190 mm**	
Initial concrete temperature: **23.2 °C**		Final concrete temperature: **25.1 °C**	
Concrete Age (days)	Specific Gravity (kg/m3)	Compressive Strength (MPa)	
7	2375	40.1	
7	2376	40.5	
28	2381	46.1	
28	2372	46.8	
Mean value for specific gravity (kg/m3)	**2376**	Mean value for compressive strength (MPa)	**46.4**

Table 6.25: **Laboratory test results for C40.3 made with company (A) cement**

Initial slump: **195 mm**		Final slump (After 30 minutes): **155 mm**	
Initial concrete temperature: **22.2 °C**		Final concrete temperature: **24.1 °C**	
Concrete Age (days)	Specific Gravity (kg/m3)		Compressive Strength (MPa)
7	2411		37.9
7	2416		38.1
28	2410		45.5
28	2409		44.9
Mean value for specific gravity (kg/m3)	2411	Mean value for compressive strength (MPa)	45.2

Table 6.26: **Laboratory test results for C40.3 made with company (B) cement**

Initial slump: **205 mm**		Final slump (After 30 minutes): **150 mm**	
Initial concrete temperature: **24.9 °C**		Final concrete temperature: **26.1 °C**	
Concrete Age (days)	Specific Gravity (kg/m3)		Compressive Strength (MPa)
7	2411		36.8
7	2409		37.5
28	2405		45.8
28	2409		45.1
Mean value for specific gravity (kg/m3)	2408	Mean value for compressive strength (MPa)	45.5

Table 6.27: **Laboratory test results for C40.4 made with company (A) cement**

Initial slump: **195 mm**		Final slump (After 30 minutes): **145 mm**	
Initial concrete temperature: **21.9 °C**		Final concrete temperature: **23.2 °C**	
Concrete Age (days)	Specific Gravity (kg/m3)		Compressive Strength (MPa)
7	2385		37.1
7	2381		36.9
28	2386		45.9
28	2388		46.3
Mean value for specific gravity (kg/m3)	2385	Mean value for compressive strength (MPa)	46.1

Table 6.28: **Laboratory test results for C40.4 made with company (B) cement**

| Initial slump: **190 mm** | | Final slump (After 30 minutes): **150 mm** | |
| Initial concrete temperature: **21.1 °C** | | Final concrete temperature: **23.5 °C** | |
Concrete Age (days)	Specific Gravity (kg/m3)	Compressive Strength (MPa)	
7	2375	36.5	
7	2371	36.1	
28	2381	44.5	
28	2379	45.3	
Mean value for specific gravity (kg/m3)	**2376**	Mean value for compressive strength (MPa)	**44.9**

Table 6.29: **Laboratory test results for C45.1 made with company (A) cement**

| Initial slump: **210 mm** | | Final slump (After 30 minutes): **195 mm** | |
| Initial concrete temperature: **23.5 °C** | | Final concrete temperature: **24.9 °C** | |
Concrete Age (days)	Specific Gravity (kg/m3)	Compressive Strength (MPa)	
7	2401	44.5	
7	2405	45.3	
28	2399	50.1	
28	2403	51.1	
Mean value for specific gravity (kg/m3)	**2402**	Mean value for compressive strength (MPa)	**50.6**

Table 6.30: **Laboratory test results for C45.1 made with company (B) cement**

| Initial slump: **205 mm** | | Final slump (After 30 minutes): **200 mm** | |
| Initial concrete temperature: **21.1 °C** | | Final concrete temperature: **23.1 °C** | |
Concrete Age (days)	Specific Gravity (kg/m3)	Compressive Strength (MPa)	
7	2381	45.5	
7	2382	46.2	
28	2385	51.1	
28	2379	51.5	
Mean value for specific gravity (kg/m3)	**2382**	Mean value for compressive strength (MPa)	**51.3**

Table 6.31: **Laboratory test results for C45.2 made with company (A) cement**

Initial slump: **200 mm**		Final slump (After 30 minutes): **170 mm**	
Initial concrete temperature: **21.5 °C**		Final concrete temperature: **23.1 °C**	
Concrete Age (days)	Specific Gravity (kg/m3)	Compressive Strength (MPa)	
7	2399	43.1	
7	2391	42.9	
28	2398	50.1	
28	2389	50.8	
Mean value for specific gravity (kg/m3)	2394	Mean value for compressive strength (MPa)	**50.4**

Table 6.32: **Laboratory test results for C45.2 made with company (B) cement**

Initial slump: **205 mm**		Final slump (After 30 minutes): **165 mm**	
Initial concrete temperature: **22.5 °C**		Final concrete temperature: **24.5 °C**	
Concrete Age (days)	Specific Gravity (kg/m3)	Compressive Strength (MPa)	
7	2381	42.9	
7	2379	43.6	
28	2385	50.8	
28	2381	49.5	
Mean value for specific gravity (kg/m3)	2381	Mean value for compressive strength (MPa)	**50.1**

Table 6.33: **Laboratory test results for C50.1 made with company (A) cement**

Initial slump: **210 mm**		Final slump (After 30 minutes): **210 mm**	
Initial concrete temperature: **21.1 °C**		Final concrete temperature: **22.6 °C**	
Concrete Age (days)	Specific Gravity (kg/m3)	Compressive Strength (MPa)	
7	2409	48.5	
7	2406	49.5	
28	2408	57.2	
28	2406	56.6	
Mean value for specific gravity (kg/m3)	2407	Mean value for compressive strength (MPa)	**56.9**

Table 6.34: Laboratory test results for C50.1 made with company (B) cement

Initial slump: **205 mm** | Final slump (After 30 minutes): **195 mm**
Initial concrete temperature: **24.1 °C** | Final concrete temperature: **25.9 °C**

Concrete Age (days)	Specific Gravity (kg/m3)	Compressive Strength (MPa)
7	2401	49.5
7	2396	48.7
28	2402	57.3
28	2408	57.9
Mean value for specific gravity (kg/m3)	2402	Mean value for compressive strength (MPa) = 57.6

Table 6.35: Laboratory test results for C50.2 made with company (A) cement

Initial slump: **210 mm** | Final slump (After 30 minutes): **200 mm**
Initial concrete temperature: **21.6 °C** | Final concrete temperature: **23.2 °C**

Concrete Age (days)	Specific Gravity (kg/m3)	Compressive Strength (MPa)
7	2405	48.5
7	2401	49.2
28	2408	57.3
28	2401	57.5
Mean value for specific gravity (kg/m3)	2403	Mean value for compressive strength (MPa) = 57.4

Table 6.36: Laboratory test results for C50.2 made with company (B) cement

Initial slump: **205 mm** | Final slump (After 30 minutes): **205 mm**
Initial concrete temperature: **21.1 °C** | Final concrete temperature: **23.1 °C**

Concrete Age (days)	Specific Gravity (kg/m3)	Compressive Strength (MPa)
7	2401	49.9
7	2399	49.3
28	2406	57.3
28	2401	56.9
Mean value for specific gravity (kg/m3)	2402	Mean value for compressive strength (MPa) = 57.1

Materials Research Forum LLC
https://doi.org/10.21741/9781644900598

Table 6.37: Laboratory test results for C60 made with company (A) cement

Concrete Age (days)	Specific Gravity (kg/m3)	Compressive Strength (MPa)
Initial slump: **215 mm**		Final slump (After 30 minutes): **210 mm**
Initial concrete temperature: **22.3 °C**		Final concrete temperature: **24.1 °C**
7	2410	59.1
7	2415	58.5
28	2411	67.9
28	2413	68.5
Mean value for specific gravity (kg/m3)	2412	Mean value for compressive strength (MPa) 68.2

Table 6.38: Laboratory test results for C60 made with company (B) cement

Concrete Age (days)	Specific Gravity (kg/m3)	Compressive Strength (MPa)
Initial slump: **210 mm**		Final slump (After 30 minutes): **195 mm**
Initial concrete temperature: **22.5 °C**		Final concrete temperature: **24.1 °C**
7	2402	60.1
7	2406	59.9
28	2410	68.8
28	2406	67.5
Mean value for specific gravity (kg/m3)	2406	Mean value for compressive strength (MPa) 68.1

Table 6.39: Laboratory test results for C70 made with company (A) cement

Concrete Age (days)	Specific Gravity (kg/m3)	Compressive Strength (MPa)
Initial slump: **220 mm**		Final slump (After 30 minutes): **220 mm**
Initial concrete temperature: **22.9 °C**		Final concrete temperature: **24.1 °C**
7	2420	66.5
7	2416	65.3
28	2418	77.5
28	2420	78.9
Mean value for specific gravity (kg/m3)	2418	Mean value for compressive strength (MPa) 78.2

Table 6.40: **Laboratory test results for C70 made with company (B) cement**		
Initial slump: **210 mm**	Final slump (After 30 minutes): **210 mm**	
Initial concrete temperature: **21.0 °C**	Final concrete temperature: **22.9 °C**	
Concrete Age (days)	Specific Gravity (kg/m3)	Compressive Strength (MPa)
7	2408	67.5
7	2411	68.8
28	2409	79.5
28	2418	79.1
Mean value for specific gravity (kg/m3)	**2411**	Mean value for compressive strength (MPa) : **79.3**

6.2   Plant tests for checking concrete mix design

Finally, we tested our mix designs in the real plant and for real projects. First, we produced the concrete with defined specifications in the batching plant and put it in the truck mixer. Then we got a sample of the concrete from the truck mixer and mold it. Finally, we tested the cubes or cylinders at the age of seven and twenty-eight days.

For C25 to C45 concretes, we used 15x15 cube specimens but we converted the results to the cylinder and for C50 to C70 concretes, we used 15x30 standard cylinder specimens. So, there is no need of conversion for these specimens.

6.2.1   Test results for plant

You can see the plant test results for one year at these figures:

Day	Concrete Type	Batching No	Slump (mm)	7-day-Comp Strngth Specimen1 (Mpa)	7-day-Comp Strngth Specimen2 (Mpa)	7-day-Comp Strength Mean (Mpa)	28-day-Comp Strngth Specimen1 (Mpa)	28-day-Comp Strngth Specimen2 (Mpa)	28-day-Comp Strength Mean (Mpa)
					November		2016		
1	C25.2	2	180	25.1	25.3	25.2	30.1	30	30.1
2	C40.1	1	220	39.9	38.5	39.2	44.2	43.8	44.0
3	C35.2	2	180	32.7	32.2	32.5	39.3	39.3	39.3
4	C30.1	1	190	31.5	30.9	31.2	36.2	34.2	35.2
5	C30.1	2	170	31	31.2	31.1	33.1	36.9	35.0
6	C35.1	1	180	35.7	34.4	35.1	42.1	40.8	41.5
7	C45.2	1	200	43.5	44.8	44.2	51.1	50.2	50.7
8	C70	2	220	59.5	58.7	59.1	77.9	79	78.5
9	C40.2	1	190	38.5	39.2	38.9	44.2	45.3	44.8
10	C25.2	2	180	24.2	26.1	25.2	31.1	32	31.6
11	C30.3	1	150	32.6	31.7	32.2	39.1	37.9	38.5
12	C30.3	2	180	29.3	29.1	29.2	34.7	34.8	34.8
13	C60	1	210	50.2	49.8	50.0	66.5	67.8	67.2
14	C50.2	2	210	44.8	43.5	44.2	58.5	60.2	59.4
15	C35.4	2	180	34.4	35.6	35.0	43.2	41.8	42.5
16	C25.1	1	150	29.3	29.6	29.5	37.1	35.8	36.5
17	C25.1	2	180	21.4	21.5	21.5	28.5	28.3	28.4
18	C30.4	1	190	28.5	29.5	29.0	34.8	37.5	36.2
19	C25.2	2	160	23.4	25.3	24.4	32.1	31.7	31.9
20	C35.1	1	150	39	39.2	39.1	43.5	41.7	42.6
21	C70	2	210	55.6	57.9	56.8	78.3	79.6	79.0
22	C25.1	2	190	28.1	26.6	27.4	34.9	33.9	34.4
23	C30.2	1	180	34.4	33.1	33.8	40	40.9	40.5
24	C30.4	2	150	31.5	33.8	32.7	42.7	42.6	42.7
25	C25.2	2	200	22.2	21	21.6	27.8	29.2	28.5
26	C30.1	1	140	31.6	32.6	32.1	37.3	39.7	38.5
27	C45.2	2	190	43.3	44.2	43.8	49.5	50.3	49.9
28	C40.3	1	180	39.5	38.8	39.2	45.5	46.3	45.9
29	C60	2	210	50.3	51.2	50.8	66.3	67.1	66.7
30	C60	1	200	51.5	52	51.8	66.5	67.3	66.9

Figure 6.5: Plant test results for November 2016

147

Day	Concrete Type	Batching No	Slump (mm)	December	2016				
				7-day-Comp Strngth Specimen1 (Mpa)	7-day-Comp Strngth Specimen2 (Mpa)	7-day-Comp Strength Mean (Mpa)	28-day-Comp Strngth Specimen1 (Mpa)	28-day-Comp Strngth Specimen2 (Mpa)	28-day-Comp Strength Mean (Mpa)
1	C30.1	2	180	30.5	29.9	30.2	38.2	38.9	38.6
2	C25.2	1	180	24.9	26.2	25.6	35	37.3	36.2
3	C35.3	2	170	38.1	38.3	38.2	52.1	50.2	51.2
4	C70	2	200	61.1	62.5	61.8	80.3	81.5	80.9
5	C45.2	1	190	38.8	37.5	38.2	50.3	51.5	50.9
6	C30.1	2	170	32.8	32.6	32.7	41.8	39.7	40.8
7	C60	2	200	55.5	56.2	55.9	66.2	67.3	66.8
8	C70	1	220	59.8	61.3	60.6	76.6	77.8	77.2
9	C25.2	1	180	24.8	23.9	24.4	34.9	34.8	34.9
10	C45.1	2	180	43.5	44.1	43.8	49.8	48.5	49.2
11	C30.4	2	180	24.3	25.1	24.7	34	34.5	34.3
12	C35.3	2	190	31.2	32.2	31.7	39.8	40.3	40.1
13	C30.4	1	180	28.5	28.1	28.3	35.5	35.9	35.7
14	C30.3	2	180	29.4	29.1	29.3	41.4	41.8	41.6
15	C30.3	1	170	26.1	26.5	26.3	34.8	35.9	35.4
16	C35.3	2	180	35.8	35.9	35.9	43.3	42.8	43.1
17	C25.2	2	180	26.3	24.9	25.6	32.5	32.5	32.5
18	C45.2	1	220	43.3	45.3	44.3	53.4	50.9	52.2
19	C45.1	2	200	44.3	43.5	43.9	51.5	50.8	51.2
20	C25.2	2	170	23	24.7	23.9	33.2	33.2	33.2
21	C40.4	1	180	34.5	34.2	34.4	44.8	44.7	44.8
22	C25.1	2	180	20.5	19.5	20.0	30.3	31.7	31.0
23	C50.1	1	220	45.5	46.3	45.9	58.5	48.6	53.6
24	C35.4	2	180	33.4	32.7	33.1	43.9	42.1	43.0
25	C40.4	1	220	38	37.9	38.0	48.2	49.7	49.0
26	C60	2	220	53.5	52.8	53.2	66.6	67	66.8
27	C45.2	1	190	41	42.5	41.8	50	49.5	49.8
28	C35.3	2	200	29.3	29.5	29.4	42.7	42.8	42.8
29	C35.3	1	180	33.7	32.6	33.2	45.1	48.4	46.8
30	C35.3	1	190	29.4	29.4	29.4	42.4	42.4	42.4
31	C60	2	200	55.3	54.5	54.9	66.3	67.7	67.0

Figure 6.6: Plant test results for December 2016

Day	Concrete Type	Batching No	Slump (mm)	7-day-Comp Strngth Specimen1 (Mpa)	7-day-Comp Strngth Specimen2 (Mpa)	7-day-Comp Strength Mean (Mpa)	28-day-Comp Strngth Specimen1 (Mpa)	28-day-Comp Strngth Specimen2 (Mpa)	28-day-Comp Strength Mean (Mpa)
					January		2017		
1	C35.3	1	180	26.6	25.9	26.3	39.3	39.7	39.5
2	C40.4	1	180	37.2	37	37.1	48.6	48.1	48.4
3	C60	2	190	53.2	54	53.6	64.5	66.3	65.4
4	C45.2	2	180	37.1	38.5	37.8	50.2	48.5	49.4
5	C40.3	2	220	36.3	36.7	36.5	46	47.9	47.0
6	C25.2	1	180	21.8	22.2	22.0	32	32.6	32.3
7	C30.1	2	190	23.5	23	23.3	33.1	32.6	32.9
8	C45.1	1	180	44.7	43.8	44.3	48.5	51	49.8
9	C30.1	2	180	28.8	27.9	28.4	34.3	35.9	35.1
10	C70	2	200	60.1	59.4	59.8	77.9	79.3	78.6
11	C50.1	2	190	44.3	46.9	45.6	59.6	60.3	60.0
12	C35.1	2	190	31.1	31	31.1	40.4	38.7	39.6
13	C30.2	1	180	25.5	26.8	26.2	33.6	34	33.8
14	C35.2	2	190	32.3	34.9	33.6	42.2	41.9	42.1
15	C25.2	2	18	21.1	19.1	20.1	31.6	31.2	31.4
16	C35.3	1	190	32.8	32.2	32.5	40.3	42.3	41.3
17	C60	2	200	54.3	53.9	54.1	68.8	69	68.9
18	C30.2	2	180	28.7	27	27.9	38.6	41.3	40.0
19	C35.1	1	190	27.9	30.6	29.3	39.1	39.4	39.3
20	C30.1	2	180	28.8	27.1	28.0	34.5	40.9	37.7
21	C35.1	1	180	32	32	32.0	41.5	41.5	41.5
22	C70	1	220	69.9	68.7	69.3	82.2	83.9	83.1
23	C30.1	2	190	27.7	27.6	27.7	35.9	39.2	37.6
24	C25.1	1	180	24.9	23.2	24.1	32	29.9	31.0
25	C30.4	2	190	25.6	25.1	25.4	35.3	35.4	35.4
26	C30.3	1	170	31.8	31.6	31.7	41	43.2	42.1
27	C40.4	1	210	43.2	43.5	43.4	56.1	55.5	55.8
28	C50.1	2	220	50.6	49.6	50.1	56.7	59.5	58.1
29	C45.2	2	190	40.5	41.1	40.8	49.9	50.6	50.3
30	C50.2	2	190	47.9	47.1	47.5	55.9	59.9	57.9
31	C45.1	1	190	40.6	41.9	41.3	49.9	51.3	50.6

Figure 6.7: Plant test results for January 2017

Materials Research Forum LLC

https://doi.org/10.21741/9781644900598

Day	Concrete Type	Batching No	Slump (mm)	February			2017		
				7-day-Comp Strngth Specimen1 (Mpa)	7-day-Comp Strngth Specimen2 (Mpa)	7-day-Comp Strength Mean (Mpa)	28-day-Comp Strngth Specimen1 (Mpa)	28-day-Comp Strngth Specimen2 (Mpa)	28-day-Comp Strength Mean (Mpa)
1	C60	1	220	53.5	52.8	53.2	66.3	65.5	65.9
2	C30.1	1	200	26.3	25.4	25.9	36.1	36.5	36.3
3	C45.2	2	190	42.2	41.8	42.0	48.5	47.9	48.2
4	C40.3	2	180	32.8	36.5	34.7	44	45.4	44.7
5	C25.2	1	180	23.5	22.7	23.1	28.8	29.6	29.2
6	C35.4	2	180	26.4	27	26.7	40.2	37.9	39.1
7	C35.1	1	180	35.3	35.2	35.3	43.3	46.2	44.8
8	C35.2	2	170	30.9	30.9	30.9	41	41	41.0
9	C25.2	2	180	21.4	18.4	19.9	30.9	31.7	31.3
10	C35.3	1	190	31.4	31.9	31.7	43	41.6	42.3
11	C30.3	2	170	25.7	25.5	25.6	37.2	39	38.1
12	C35.4	1	180	29.8	31.5	30.7	41.1	43.5	42.3
13	C40.4	2	200	41.2	36	38.6	53	51.3	52.2
14	C50.1	1	200	45.1	44.2	44.7	56.2	55.3	55.8
15	C70	2	210	60.2	59.3	59.8	76.6	78	77.3
16	C30.3	1	160	24.3	24.1	24.2	36.2	35.3	35.8
17	C50.1	2	220	46.4	45	45.7	55.5	54.7	55.1
18	C35.4	1	180	22.8	23	22.9	35.2	34.8	35.0
19	C25.2	2	190	22	21.7	21.9	31.1	33.3	32.2
20	C40.3	1	180	36.4	35.9	36.2	46.7	47.2	47.0
21	C35.4	2	170	33.7	34	33.9	44.7	42.4	43.6
22	C35.2	1	190	32.8	35.1	34.0	42.8	42.5	42.7
23	C70	2	200	62.3	65.9	64.1	79.5	80.8	80.2
24	C30.1	1	170	28.1	28.8	28.5	37.2	32.7	35.0
25	C35.3	2	180	36	36.8	36.4	43.6	42.3	43.0
26	C35.1	1	190	33.2	33.6	33.4	39	37.3	38.2
27	C30.2	2	190	26.6	27.2	26.9	37	34.4	35.7
28	C35.2	1	180	33.4	30.5	32.0	42.4	42.8	42.6
29	C50.1	1	200	43.6	44	43.8	57.9	58.6	58.3

Figure 6.8: Plant test results for February 2017

Day	Concrete Type	Batching No	Slump (mm)	7-day-Comp Strngth Specimen1 (Mpa)	7-day-Comp Strngth Specimen2 (Mpa)	7-day-Comp Strength Mean (Mpa)	28-day-Comp Strngth Specimen1 (Mpa)	28-day-Comp Strngth Specimen2 (Mpa)	28-day-Comp Strength Mean (Mpa)
					March		2017		
1	C40.4	2	180	36.3	37.9	37.1	50.1	47.6	48.9
2	C30.3	1	190	27.3	27	27.2	37.2	40.9	39.1
3	C35.2	2	190	35.7	34.9	35.3	46.1	43.6	44.9
4	C40.4	1	180	37.8	36.3	37.1	45.6	48.5	47.1
5	C35.1	2	180	34.8	35.2	35.0	44.7	46.9	45.8
6	C60	1	200	53.3	52.5	52.9	66.5	65.7	66.1
7	C30.1	1	170	31.6	25.7	28.7	39.5	33.3	36.4
8	C30.1	2	180	30.5	29.1	29.8	41.4	41.3	41.4
9	C35.3	2	180	32.9	33.2	33.1	40.8	40.3	40.6
10	C35.3	1	170	33.1	34.2	33.7	41.3	41	41.2
11	C25.2	2	160	22.5	23.8	23.2	29.1	27.6	28.4
12	C70	1	190	65.5	67.1	66.3	81.5	82.6	82.1
13	C50.1	1	190	48.3	48.3	48.3	60	60	60.0
14	C30.1	2	170	30.2	31.8	31.0	38.7	39.7	39.2
15	C25.2	1	170	22.5	22.1	22.3	30	31.1	30.6
16	C40.1	2	170	37.2	37.7	37.5	48.2	46.5	47.4
17	C30.1	2	180	30.1	30.2	30.2	36.9	38.8	37.9
18	C30.1	1	180	27.7	27.5	27.6	33.5	35.7	34.6
19	C35.1	2	170	26.9	30	28.5	38.5	39.5	39.0
20	0	0	0	0	0	0.0	0	0	0.0
21	0	0	0	0	0	0.0	0	0	0.0
22	0	0	0	0	0	0.0	0	0	0.0
23	0	0	0	0	0	0.0	0	0	0.0
24	0	0	0	0	0	0.0	0	0	0.0
25	0	0	0	0	0	0.0	0	0	0.0
26	C45.2	2	200	49.3	47.5	48.4	50.7	50.7	50.7
27	C35.2	2	180	33.7	34.1	33.9	41.6	41.7	41.7
28	C25.2	1	180	24.4	23.9	24.2	30.1	31.2	30.7
29	C45.2	2	190	42.2	41	41.6	49.6	50.8	50.2
30	C30.3	2	180	27.5	26.3	26.9	36.8	37.9	37.4
31	C30.1	2	190	29.5	30.2	29.9	36.8	35.1	36.0

Figure 6.9: Plant test results for March 2017

Day	Concrete Type	Batching No	Slump (mm)	April			2017		
				7-day-Comp Strngth Specimen1 (Mpa)	7-day-Comp Strngth Specimen2 (Mpa)	7-day-Comp Strength Mean (Mpa)	28-days-Comp Strngth Specimen1 (Mpa)	28-days-Comp Strngth Specimen2 (Mpa)	28-day-Comp Strength Mean (Mpa)
1	C45.1	1	180	39.5	38.7	39.1	49.9	50.3	50.1
2	C30.1	1	190	29.5	28.8	29.2	35.4	34.6	35.0
3	C40.2	2	190	39.5	39.9	39.7	46.8	45.3	46.1
4	C30.2	1	200	28.8	27.9	28.4	34.5	34	34.3
5	C45.1	2	220	41.5	42.8	42.2	50.6	51.9	51.3
6	C50.2	2	210	47.3	45.9	46.6	59.9	57.8	58.9
7	C30.4	2	180	25.5	24.8	25.2	36.9	35.4	36.2
8	C30.1	1	170	30.5	31.9	31.2	37.6	36.4	37.0
9	C35.3	2	190	26.1	26.1	26.1	39.9	40	40.0
10	C60	1	190	54.6	53.9	54.3	66.6	67.9	67.3
11	C25.1	1	190	25.5	24.9	25.2	31.1	32	31.6
12	C45.2	2	200	40.3	41.9	41.1	50.8	49.5	50.2
13	C30.2	2	190	29.5	30.8	30.2	35	36.7	35.9
14	C35.1	2	170	34.1	34.1	34.1	41.8	43.1	42.5
15	C30.1	2	170	26.7	26.8	26.8	33.9	35.2	34.6
16	C30.1	1	190	25.4	20.5	23.0	33.1	34.3	33.7
17	C25.2	2	180	23	22.5	22.8	30.6	30.1	30.4
18	C45.2	2	180	42.5	43	42.8	47.8	48.9	48.4
19	C30.2	1	180	27	26.7	26.9	33.7	33.6	33.7
20	C30.1	2	180	29.8	28.6	29.2	35.7	34.6	35.2
21	C35.1	1	160	35.3	37	36.2	40.6	44.7	42.7
22	C70	2	200	59.5	58.8	59.2	73.3	74.9	74.1
23	C25.2	2	180	22.8	23.5	23.2	28.9	29.1	29.0
24	C25.2	2	170	23.7	23.3	23.5	29.5	29.8	29.7
25	C35.3	1	190	31.6	32.9	32.3	37.7	38.5	38.1
26	C35.1	1	160	31.4	32	31.7	37.3	38.8	38.1
27	C30.1	2	190	25.9	25.7	25.8	32	32.8	32.4
28	C25.2	1	180	21.1	20.8	21.0	26.9	27.8	27.4
29	C35.3	2	170	36.2	33.1	34.7	42.1	44.1	43.1
30	C30.4	1	170	28.7	29.2	29.0	37.5	34.1	35.8

Figure 6.10: Plant test results for April 2017

| Day | Concrete Type | Batching No | Slump (mm) | May | 2017 | | | | |
				7-day-Comp Strngth Specimen1 (Mpa)	7-day-Comp Strngth Specimen2 (Mpa)	7-day-Comp Strength Mean (Mpa)	28-day-Comp Strngth Specimen1 (Mpa)	28-day-Comp Strngth Specimen2 (Mpa)	28-day-Comp Strength Mean (Mpa)
1	C45.1	1	190	43.5	41.1	42.3	50.2	49.3	49.8
2	C30.1	1	190	28.8	31.1	30.0	38	37.7	37.9
3	C35.3	2	180	32.1	31.6	31.9	40.1	39.8	40.0
4	C25.1	2	170	25.7	22.5	24.1	32.3	31.6	32.0
5	C60	1	220	54.9	53.5	54.2	66.6	68.2	67.4
6	C30.1	2	160	32.5	32.5	32.5	38.9	38.8	38.9
7	C30.2	2	180	28.1	27.7	27.9	36	34.4	35.2
8	C25.2	1	180	22.9	23.6	23.3	29.5	30.8	30.2
9	C25.2	2	170	25.8	25.6	25.7	32.2	32	32.1
10	C40.3	2	190	41.2	41.2	41.2	47.6	49.6	48.6
11	C25.2	1	170	25.5	23.3	24.4	33.9	32.9	33.4
12	C50.1	2	200	43.5	44.9	44.2	56.4	55.9	56.2
13	C45.1	1	220	41.5	42.4	42.0	48.6	49.5	49.1
14	C25.2	2	160	23.9	25	24.5	32.1	31	31.6
15	C70	1	210	61.9	60.7	61.3	77.9	79.9	78.9
16	C35.4	2	180	35.1	35.8	35.5	44.1	46	45.1
17	C35.4	2	170	32.8	34.5	33.7	44	43.1	43.6
18	C35.4	1	190	33.5	34	33.8	42.1	41	41.6
19	C40.3	2	190	35.2	36.9	36.1	45.1	44	44.6
20	C25.2	2	170	20.4	23.5	22.0	32.6	32.7	32.7
21	C25.1	1	180	20.1	20.5	20.3	27.6	26.7	27.2
22	C45.2	2	190	41.1	40.8	41.0	48.9	49.7	49.3
23	C50.1	2	180	47.4	45.2	46.3	56.4	55.4	55.9
24	C30.1	1	180	25.9	26.4	26.2	34.3	33.8	34.1
25	C30.2	2	180	26.4	27.5	27.0	34.5	35.5	35.0
26	C30.3	1	180	29.5	31.4	30.5	39.6	40	39.8
27	C30.3	2	190	30.8	31.7	31.3	37	37.7	37.4
28	C30.2	1	180	28.2	27.2	27.7	34.8	35.5	35.2
29	C50.2	2	210	44.5	46.1	45.3	59.8	60.3	60.1
30	C35.3	2	170	35.5	36.3	35.9	46.7	44.1	45.4
31	C25.2	1	190	23.8	23.6	23.7	31.3	30.7	31.0

Figure 6.11: Plant test results for May 2017

					June		2017		
Day	Concrete Type	Batching No	Slump (mm)	7-day-Comp Strngth Specimen1 (Mpa)	7-day-Comp Strngth Specimen2 (Mpa)	7-day-Comp Strength Mean (Mpa)	28-day-Comp Strngth Specimen1 (Mpa)	28-day-Comp Strngth Specimen2 (Mpa)	28-day-Comp Strength Mean (Mpa)
1	C30.4	2	190	31.9	32.2	32.1	38.8	40	39.4
2	C30.2	2	180	27.4	27.1	27.3	34.2	34.9	34.6
3	C50.1	1	220	54.4	52.8	53.6	61.3	63.6	62.5
4	C30.3	2	180	33	31.4	32.2	39.6	40	39.8
5	C45.1	2	190	40.3	39.8	40.1	49.5	51.3	50.4
6	C35.4	1	190	33.2	33.4	33.3	40.5	42.2	41.4
7	C50.2	2	190	52.2	53.1	52.7	63.7	64.2	64.0
8	C25.2	1	200	22.1	23.2	22.7	28.9	29.5	29.2
9	C30.1	2	180	29.1	25.8	27.5	37.2	38.1	37.7
10	C25.1	1	170	22.5	22.3	22.4	29.8	29.2	29.5
11	C35.1	2	180	34.6	32.9	33.8	42	41.3	41.7
12	C70	1	220	61.3	62.9	62.1	77.6	76.1	76.9
13	C35.4	2	170	36.3	36.6	36.5	43.1	43.8	43.5
14	C45.1	1	180	43.9	44	44.0	51.2	53.6	52.4
15	C50.1	2	190	45.5	43.1	44.3	56.6	57.9	57.3
16	C25.2	1	190	25.3	28.9	27.1	34.3	32.7	33.5
17	C30.1	2	170	28.5	27	27.8	34.1	35	34.6
18	C30.1	2	190	26.9	27.3	27.1	33.5	34.9	34.2
19	C70	2	200	61.6	60.8	61.2	78.5	80.5	79.5
20	C30.3	1	180	28.3	29.9	29.1	38.7	38.4	38.6
21	C25.2	2	200	20.2	20.1	20.2	27.5	28.9	28.2
22	C25.2	1	180	21.6	22.2	21.9	28.9	30.3	29.6
23	C70	2	220	66.8	66.8	66.8	79.5	77.4	78.5
24	C30.4	2	180	26.5	26.2	26.4	35	34.3	34.7
25	C30.4	1	190	25.3	26	25.7	34.5	33.4	34.0
26	C35.2	1	180	35	35	35.0	43.4	43.4	43.4
27	C35.4	2	170	30.8	30.7	30.8	40.1	37.2	38.7
28	C60	2	220	59.8	61.3	60.6	70.9	70.3	70.6
29	C40.1	1	180	39.4	38.6	39.0	47	46.2	46.6
30	C45.1	2	200	44.2	44.5	44.4	53.6	54.2	53.9

Figure 6.12: Plant test results for June 2017

Day	Concrete Type	Batching No	Slump (mm)	July			2017		
				7-day-Comp Strngth Specimen1 (Mpa)	7-day-Comp Strngth Specimen2 (Mpa)	7-day-Comp Strength Mean (Mpa)	28-day-Comp Strngth Specimen1 (Mpa)	28-day-Comp Strngth Specimen2 (Mpa)	28-day-Comp Strength Mean (Mpa)
1	C60	2	190	60.6	59.3	60.0	70.1	72.6	71.4
2	C50.2	1	200	46.8	47.3	47.1	62.2	59.5	60.9
3	C30.1	1	180	30.1	29.4	29.8	37.5	35.9	36.7
4	C25.1	2	180	22.1	25	23.6	33.5	32	32.8
5	C45.2	2	200	43.3	42.9	43.1	51.1	52	51.6
6	C30.2	1	180	29.3	28.1	28.7	31.5	35.4	33.5
7	C25.1	1	210	21	21.3	21.2	27.1	26.6	26.9
8	C50.1	2	220	52.4	52.8	52.6	60.8	63.2	62.0
9	C35.4	2	210	31.5	33.6	32.6	42.9	41.6	42.3
10	C35.4	1	190	36.2	36.5	36.4	43.2	41.9	42.6
11	C50.1	2	220	48.2	51.6	49.9	59.3	59	59.2
12	C30.2	2	190	28.1	28.8	28.5	33.7	34.1	33.9
13	C25.2	1	180	23.8	25.5	24.7	32.2	32.8	32.5
14	C70	1	220	60.3	62.5	61.4	85.6	84.9	85.3
15	C40.2	1	180	39.4	39.5	39.5	46.3	47.1	46.7
16	C40.2	2	190	37.5	36.9	37.2	44.5	43.8	44.2
17	C30.1	1	180	29.3	29.1	29.2	36.8	37.5	37.2
18	C25.2	1	170	27.5	25.8	26.7	32.2	32	32.1
19	C30.1	2	200	27.8	29	28.4	34.8	34	34.4
20	C30.1	1	170	30	29.3	29.7	36.3	37.5	36.9
21	C35.1	2	180	36.2	35.2	35.7	40.3	43.9	42.1
22	C45.2	2	180	42.2	41	41.6	52.3	50.9	51.6
23	C60	1	210	53.5	55.9	54.7	67.9	69.1	68.5
24	C40.1	1	170	40.4	40.7	40.6	48.4	46.9	47.7
25	C25.2	2	180	25.2	25.5	25.4	31.3	31.6	31.5
26	C30.2	1	180	27.6	27.4	27.5	35.2	33.9	34.6
27	C50.2	1	190	44.5	45.9	45.2	56.6	55.1	55.9
28	C30.3	2	170	26.7	27.6	27.2	36.6	37.8	37.2
29	C30.3	1	200	27.8	29.6	28.7	38.9	37.4	38.2
30	C70	2	200	60.3	63.2	61.8	77.8	80.6	79.2
31	C40.1	1	180	38.7	39.9	39.3	44.3	46.8	45.6

Figure 6.13: Plant test results for July 2017

155

Day	Concrete Type	Batching No	Slump (mm)	7-day-Comp Strngth Specimen1 (Mpa)	7-day-Comp Strngth Specimen2 (Mpa)	7-day-Comp Strength Mean (Mpa)	28-day-Comp Strngth Specimen1 (Mpa)	28-day-Comp Strngth Specimen2 (Mpa)	28-day-Comp Strength Mean (Mpa)
				August			**2017**		
1	C25.2	2	190	22.5	22.8	22.7	27.6	30	28.8
2	C30.3	1	180	30.2	31	30.6	40.2	39.3	39.8
3	C25.2	2	200	20.7	21	20.9	27	26.4	26.7
4	C30.1	1	170	29.3	29.3	29.3	35.7	37.3	36.5
5	C30.1	2	180	27.5	28.9	28.2	35.1	34.2	34.7
6	C70	1	210	59.5	58.8	59.2	74.9	77.3	76.1
7	C25.2	1	180	24.5	25.8	25.2	31.1	31.1	31.1
8	C45.1	1	180	38.7	39.1	38.9	49.5	51.2	50.4
9	C35.2	2	170	32.8	31.5	32.2	39	40	39.5
10	C50.1	1	200	45.5	43.9	44.7	60.3	58.7	59.5
11	C35.4	2	180	33.9	35.6	34.8	43.1	43.1	43.1
12	C25.2	1	180	25.5	25.3	25.4	31.9	32.9	32.4
13	C30.2	1	180	28.8	29.6	29.2	34.5	34.6	34.6
14	C35.2	1	170	32.5	31.5	32.0	40.1	39.1	39.6
15	C30.1	2	180	28	29.9	29.0	35.2	35.1	35.2
16	C40.3	1	170	37.7	39.4	38.6	45.4	45.8	45.6
17	C35.2	2	170	35.3	35.4	35.4	40.1	43.7	41.9
18	C30.4	2	190	32.4	31.5	32.0	37.7	38.2	38.0
19	C45.1	1	180	38.5	40.6	39.6	51.5	50.9	51.2
20	C60	2	210	50.6	49.8	50.2	63.3	64.5	63.9
21	C35.3	2	180	31.1	30	30.6	40.5	37.1	38.8
22	C70	1	190	63.5	61.9	62.7	74.5	73.9	74.2
23	C30.4	2	180	32	28.8	30.4	36.1	35.8	36.0
24	C30.4	1	180	28.5	27.9	28.2	36.6	38.3	37.5
25	C30.1	2	190	30.4	30.5	30.5	36.8	36	36.4
26	C30.1	2	170	28.7	28.2	28.5	35.4	34.1	34.8
27	C50.1	2	200	42.3	44.9	43.6	56.5	57	56.8
28	C35.1	2	180	35.2	33.2	34.2	41.4	42.8	42.1
29	C25.2	1	180	26.1	26.5	26.3	33	31.9	32.5
30	C30.1	2	190	28.2	26	27.1	32.7	32.1	32.4
31	C35.1	1	190	34.5	32.9	33.7	40.1	40	40.1

Figure 6.14: Plant test results for August 2017

Day	Concrete Type	Batching No	Slump (mm)	7-day-Comp Strngth Specimen1 (Mpa)	7-day-Comp Strngth Specimen2 (Mpa)	7-day-Comp Strength Mean (Mpa)	28-day-Comp Strngth Specimen1 (Mpa)	28-day-Comp Strngth Specimen2 (Mpa)	28-day-Comp Strength Mean (Mpa)
				September		2017			
1	C25.1	1	180	22	23.5	22.8	28.5	29.8	29.2
2	C30.1	2	180	28.1	26.4	27.3	34.6	35.1	34.9
3	C45.1	1	200	40.6	41.9	41.3	53.3	55.1	54.2
4	C35.2	2	200	35.9	34.8	35.4	42.4	42.3	42.4
5	C35.3	1	180	35.5	33.1	34.3	41.4	44	42.7
6	C30.3	2	180	32	31.2	31.6	35	37.3	36.2
7	C30.3	1	190	32	31.3	31.7	38.5	38.8	38.7
8	C30.2	1	170	29.9	30.3	30.1	36.5	35.7	36.1
9	C30.1	2	190	28.8	27.8	28.3	35.3	36.3	35.8
10	C60	2	200	58.3	56.8	57.6	71.4	71.4	71.4
11	C40.4	2	170	39.6	37.7	38.7	47.2	44.5	45.9
12	C30.1	1	180	30	30.3	30.2	34	36.1	35.1
13	C35.4	2	190	34.2	34.8	34.5	42.1	41.5	41.8
14	C30.2	2	190	29.4	29.4	29.4	33.9	33.8	33.9
15	C35.3	1	180	35.1	34	34.6	43.7	43.5	43.6
16	C35.3	2	190	34.4	33.1	33.8	43.4	42.7	43.1
17	C70	2	200	61.3	60.2	60.8	76.6	78.9	77.8
18	C50.1	2	190	46.5	47	46.8	58.9	59.6	59.3
19	C25.2	1	200	25.8	25.6	25.7	31.4	30	30.7
20	C25.2	2	190	21.9	22	22.0	27.5	27.6	27.6
21	C35.1	1	170	34.5	33.7	34.1	40.6	42	41.3
22	C30.2	2	220	26.5	26.2	26.4	33.1	32.2	32.7
23	C25.2	1	180	25.4	25	25.2	32.2	32.8	32.5
24	C60	2	190	50.3	53.6	52.0	68.8	65.5	67.2
25	C45.1	2	180	39.5	37.8	38.7	48.8	49.3	49.1
26	C25.2	2	190	23.4	22.9	23.2	30.5	30.5	30.5
27	C35.3	1	180	34.1	33.4	33.8	41.4	40	40.7
28	C25.2	2	180	25.3	23.4	24.4	31.9	31.5	31.7
29	C30.4	2	170	27.5	29	28.3	36.4	36.6	36.5
30	C30.1	2	200	27.8	29.1	28.5	36.1	35.1	35.6

Figure 6.15: Plant test results for September 2017

Day	Concrete Type	Batching No	Slump (mm)	7-day-Comp Strngth Specimen1 (Mpa)	7-day-Comp Strngth Specimen2 (Mpa)	7-day-Comp Strength Mean (Mpa)	28-day-Comp Strngth Specimen1 (Mpa)	28-day-Comp Strngth Specimen2 (Mpa)	28-day-Comp Strength Mean (Mpa)
					October		2017		
1	C25.2	1	200	25.5	25.1	25.3	30.1	30.6	30.4
2	C35.2	1	180	36.2	34.6	35.4	42.5	42.7	42.6
3	C25.2	1	180	22.3	21.9	22.1	29.5	29.7	29.6
4	C45.1	2	180	37.5	36.9	37.2	48.5	47.7	48.1
5	C50.1	1	190	40.3	43.2	41.8	56.5	57.9	57.2
6	C35.4	1	180	28.9	29.5	29.2	40.1	38.9	39.5
7	C70	2	200	60.3	58	59.2	74.5	76.9	75.7
8	C40.4	2	190	33.2	31.9	32.6	45.5	46.8	46.2
9	C30.3	1	200	25.3	24.9	25.1	36.6	37.8	37.2
10	C30.1	1	190	27.5	28.7	28.1	38.8	34.9	36.9
11	C70	1	200	58.9	56.5	57.7	76.6	74.7	75.7
12	C40.2	1	190	39.5	38.7	39.1	45.3	47.1	46.2
13	C45.2	2	200	40.3	39.4	39.9	50.8	52.1	51.5
14	C70	1	190	60.3	62.5	61.4	81.8	80.7	81.3
15	C30.1	2	220	28.5	27	27.8	36.7	35.4	36.1
16	C30.1	1	190	30.2	29.5	29.9	36.7	35	35.9
17	C60	2	190	51.3	53.9	52.6	66.6	68.2	67.4
18	C25.2	2	200	22.3	23.3	22.8	30	30.3	30.2
19	C30.1	2	190	32.6	31.4	32.0	36.7	38.5	37.6
20	C35.4	1	180	29.5	30.5	30.0	39.8	42.3	41.1
21	C50.2	1	190	47.3	46.5	46.9	59.6	60.8	60.2
22	C50.2	2	200	43.2	44.5	43.9	55.9	56.8	56.4
23	C30.2	2	200	26.9	23.6	25.3	35	36.1	35.6
24	C45.1	1	200	39.5	37.3	38.4	50.3	52.9	51.6
25	C30.1	2	180	31.1	32.5	31.8	38.9	39	39.0
26	C45.2	2	220	40.3	38.5	39.4	47.8	48.9	48.4
27	C70	1	190	61.3	65.5	63.4	87.6	85.2	86.4
28	C35.4	1	200	28.8	27.6	28.2	40.2	41.6	40.9
29	C25.1	2	170	22.1	22.2	22.2	29.4	28.9	29.2
30	C30.2	1	180	28.8	29.2	29.0	35.1	36	35.6
31	C60	2	190	55.3	53.2	54.3	66.9	68.2	67.6

Figure 6.16: Plant test results for October 2017

6.2.3   Batching plant testing analysis

You can see the thorough analysis for batching plant test results in table 6.41.

Table 6.41: Complete analysis for batching plant test results

Concrete Type	Number of Specimens for 1 Year	Mean Value for Slump (mm)	Mean Value for Compressive Strength (MPa)	Standard Deviation
C25.1	13	179	30.1	2.76
C25.2	46	181	31.0	1.94
C30.1	48	182	36.1	1.97
C30.2	20	185	35.2	1.94
C30.3	19	179	38.2	1.95
C30.4	14	180	36.6	2.27
C35.1	17	177	41.4	2.13
C35.2	13	181	41.8	1.63
C35.3	22	183	42.3	2.88
C35.4	19	183	41.6	2.29
C40.1	5	184	46.3	1.51
C40.2	5	188	45.6	1.05
C40.3	7	187	46.2	1.43
C40.4	9	190	48.7	3.44
C45.1	18	190	50.8	1.61
C45.2	16	196	50.1	1.17
C50.1	17	203	58.0	2.40
C50.2	9	199	59.3	2.45
C60	20	203	67.5	1.88
C70	23	204	79.0	3.17

You can also see the total analysis in these figures:
- Figure 6.17: Slump for all types of normal concrete (C25 to C35)
- Figure 6.18: Slump for all types of high strength concrete (C40 to C70)
- Figure 6.19: Standard deviation for normal concretes (C25 to C35)
- Figure 6.20 Standard deviation for high strength concretes (C40 to C70)
- Figure 6.21: Compressive strength for C25 concretes.
- Figure 6.22: Compressive strength for C30 concretes.
- Figure 6.23: Compressive strength for C35 concretes.
- Figure 6.24: Compressive strength for C40 concretes.
- Figure 6.25: Compressive strength for C45 concretes.
- Figure 6.26: Compressive strength for C50 concretes.
- Figure 6.27: Compressive strength for C60 concretes.
- Figure 6.28: Compressive strength for C70 concretes.

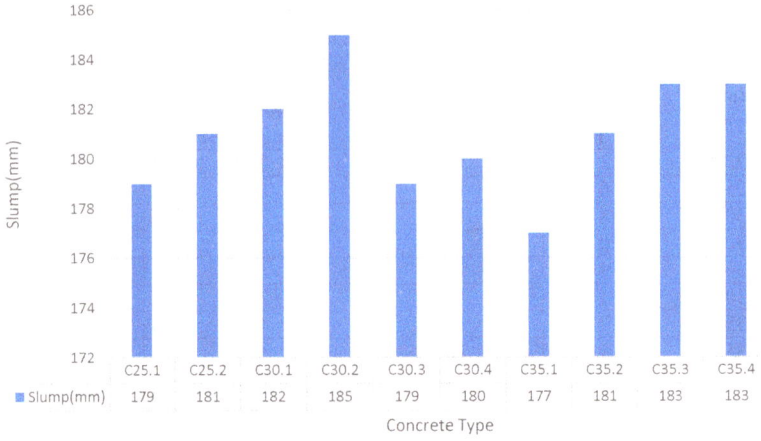

Figure(6.17): Normal concretes slump

	C25.1	C25.2	C30.1	C30.2	C30.3	C30.4	C35.1	C35.2	C35.3	C35.4
Slump(mm)	179	181	182	185	179	180	177	181	183	183

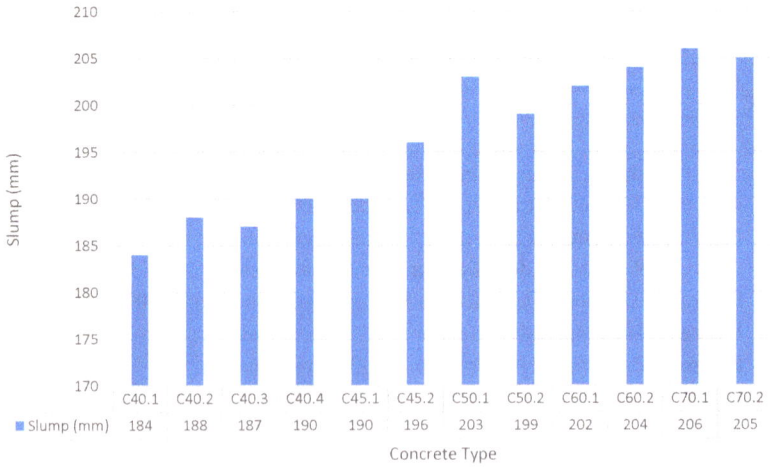

Figure(6.18): High strength concretes slump

	C40.1	C40.2	C40.3	C40.4	C45.1	C45.2	C50.1	C50.2	C60.1	C60.2	C70.1	C70.2
Slump (mm)	184	188	187	190	190	196	203	199	202	204	206	205

Figure(6.19): Standard deviation for normal concrete

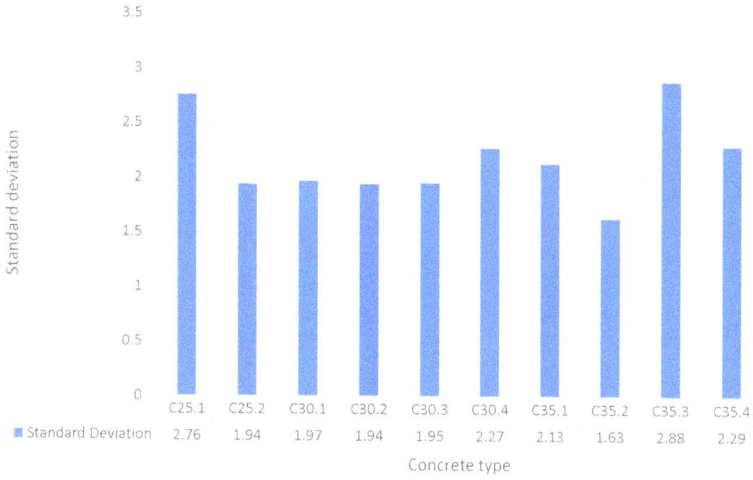

	C25.1	C25.2	C30.1	C30.2	C30.3	C30.4	C35.1	C35.2	C35.3	C35.4
Standard Deviation	2.76	1.94	1.97	1.94	1.95	2.27	2.13	1.63	2.88	2.29

Concrete type

Figure(6.20): Standard deviation for high strength concrete

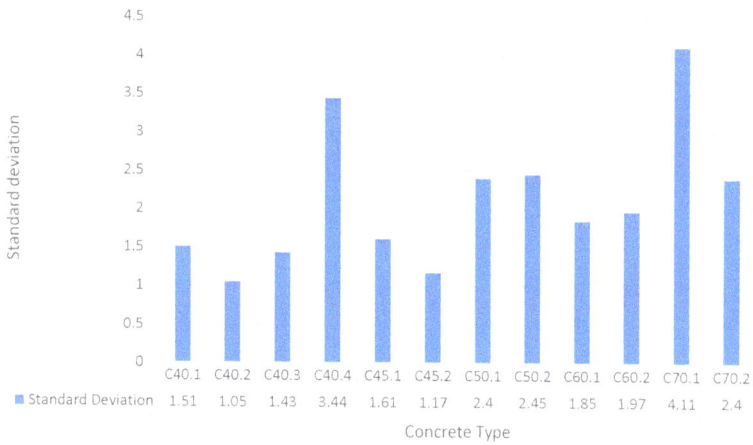

	C40.1	C40.2	C40.3	C40.4	C45.1	C45.2	C50.1	C50.2	C60.1	C60.2	C70.1	C70.2
Standard Deviation	1.51	1.05	1.43	3.44	1.61	1.17	2.4	2.45	1.85	1.97	4.11	2.4

Concrete Type

Figure(6.21): Compressive strength for C25 concrete

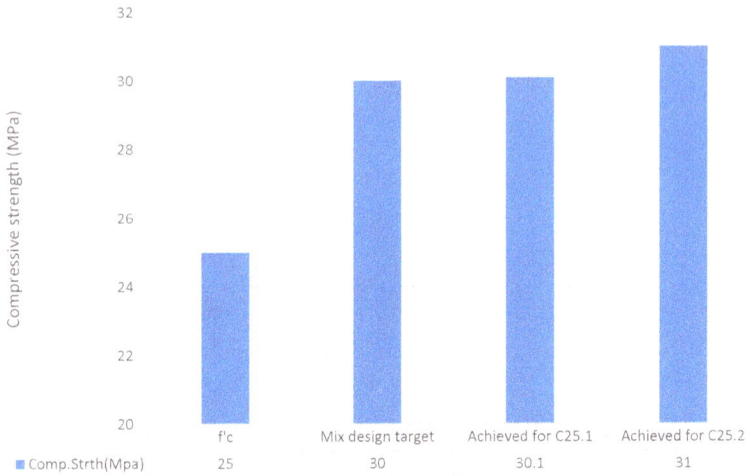

	f'c	Mix design target	Achieved for C25.1	Achieved for C25.2
■ Comp.Strth(Mpa)	25	30	30.1	31

Figure(6.22): Compressive strength for C30 concrete

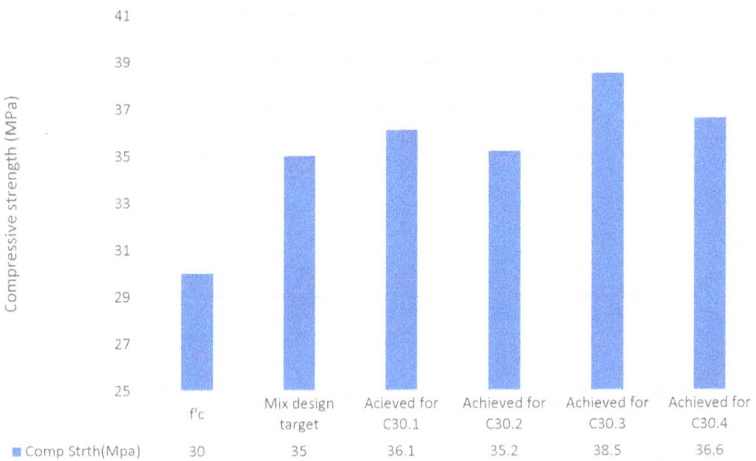

	f'c	Mix design target	Acieved for C30.1	Achieved for C30.2	Achieved for C30.3	Achieved for C30.4
■ Comp Strth(Mpa)	30	35	36.1	35.2	38.5	36.6

Figure(6.23): Compressive strength for C35 concrete

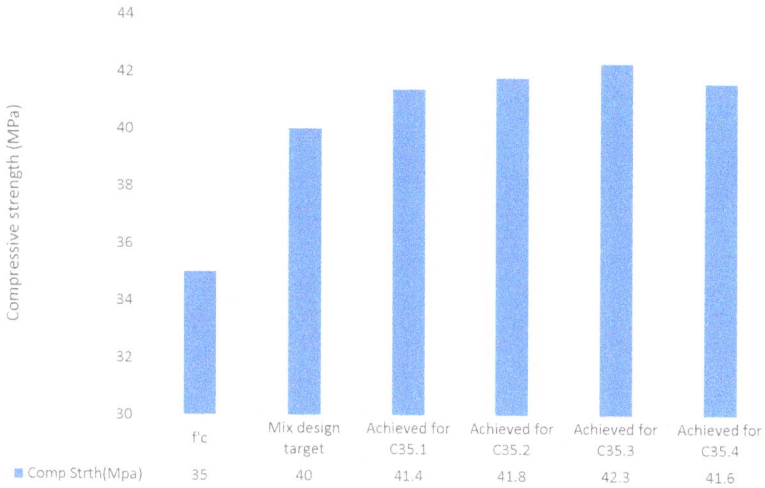

	f'c	Mix design target	Achieved for C35.1	Achieved for C35.2	Achieved for C35.3	Achieved for C35.4
Comp Strth(Mpa)	35	40	41.4	41.8	42.3	41.6

Figure(6.24): Compressive strength for C40 concrete

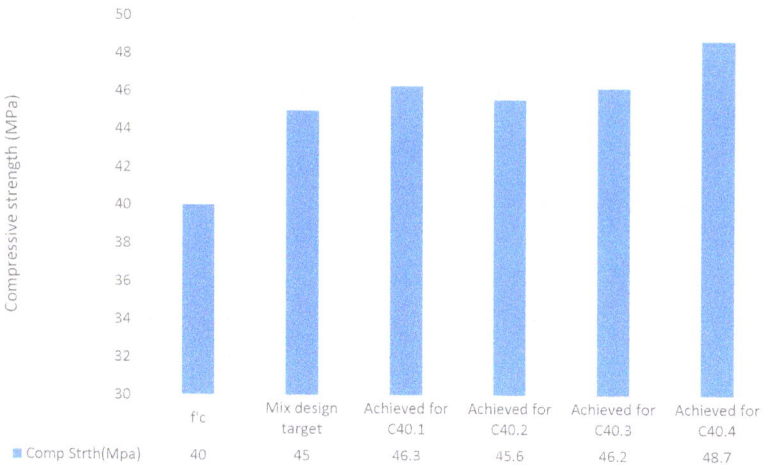

	f'c	Mix design target	Achieved for C40.1	Achieved for C40.2	Achieved for C40.3	Achieved for C40.4
Comp Strth(Mpa)	40	45	46.3	45.6	46.2	48.7

Figure (6.25): Compressive strength for C45 concrete

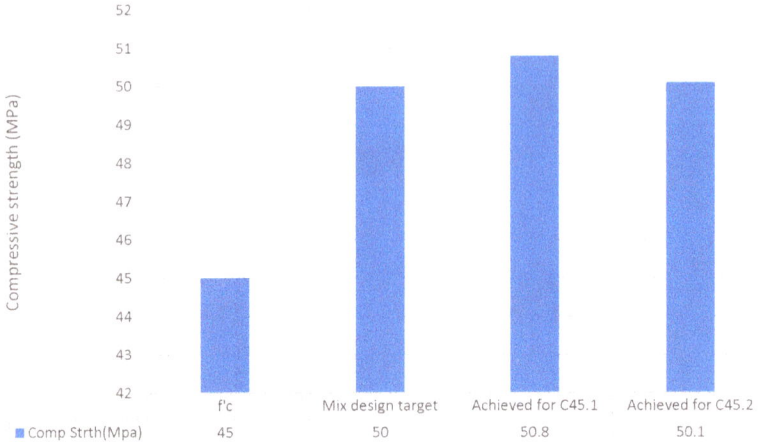

	f'c	Mix design target	Achieved for C45.1	Achieved for C45.2
Comp Strth(Mpa)	45	50	50.8	50.1

Figure (6.26): Compressive strength for C50 concrete

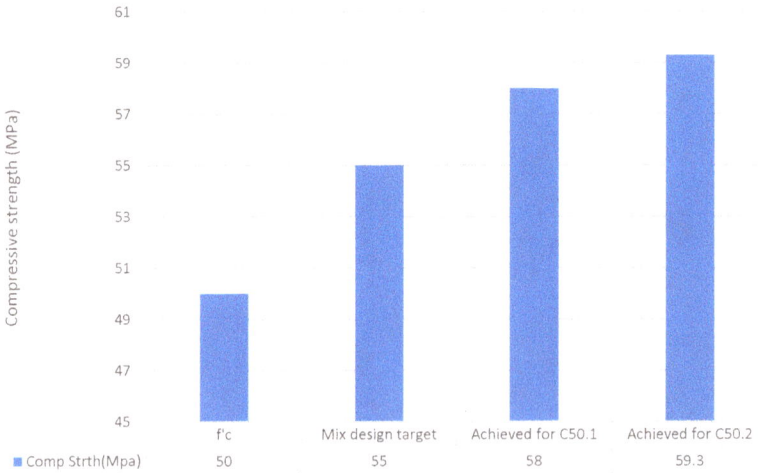

	f'c	Mix design target	Achieved for C50.1	Achieved for C50.2
Comp Strth(Mpa)	50	55	58	59.3

Figure (6.27): Compressive strength for C60 concrete

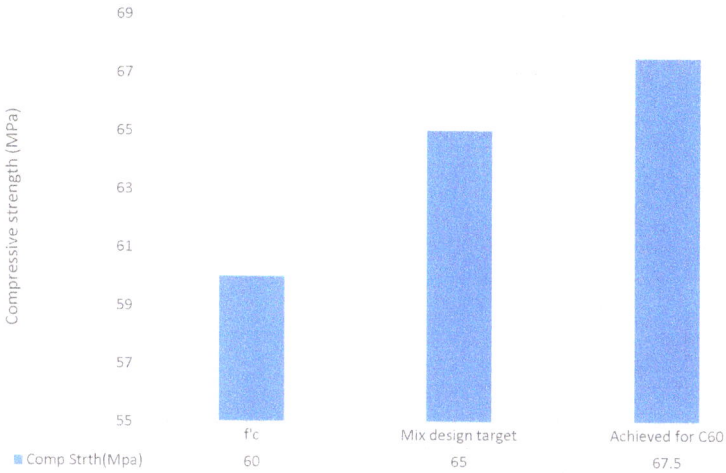

Comp Strth(Mpa)	f'c	Mix design target	Achieved for C60
	60	65	67.5

Figure (6.28): Compressive strength for C70 concrete

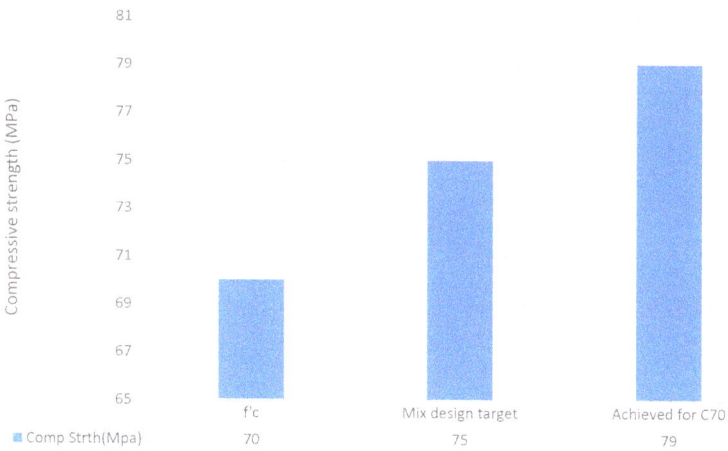

Comp Strth(Mpa)	f'c	Mix design target	Achieved for C70
	70	75	79

Conclusion

We tried to use our mix design method first in the laboratory and then in a real batching plant. The results show that our mix design method worked well.

You can see the brief test results in the table 7.1.

Table 7.1: Total test results for all types of concrete

Concrete type	Mix design target (MPa)	Lab trial company(A) cement (MPa)	Lab trial company(B) cement (MPa)	Mean value for lab trials (MPa)	Deviation from target for lab results (%)	Mean value for plant results (MPa)	Deviation from target for plant results (%)
C25.1	30	32	30.5	31.2	4.0	30.1	0.3
C25.2	30	31.2	30.8	31.0	3.3	31.0	3.3
C30.1	35	37.0	35.8	36.4	4.0	36.1	3.1
C30.2	35	35.3	35.4	35.3	0.8	35.2	0.6
C30.3	35	36.1	35.6	35.8	2.3	38.2	9.1
C30.4	35	35.8	36.6	36.2	3.4	36.6	4.6
C35.1	40	40.9	41.0	40.9	2.2	41.4	3.5
C35.2	40	40.0	41.3	40.6	1.5	41.8	4.5
C35.3	40	40.2	40.2	40.2	0.5	42.3	5.7
C35.4	40	39.8	40.6	40.2	0.5	41.6	4.0
C40.1	45	46.6	46.8	46.7	3.8	46.3	2.9
C40.2	45	46.2	46.4	46.3	2.9	45.6	1.3
C40.3	45	45.2	45.5	45.3	0.7	46.2	2.7
C40.4	45	46.1	44.9	45.5	1.1	48.7	8.2
C45.1	50	50.6	51.3	50.9	1.8	50.8	1.6
C45.2	50	50.4	50.1	50.2	0.4	50.1	0.2
C50.1	55	56.9	57.6	57.2	4.0	58.0	5.4
C50.2	55	57.4	57.1	57.2	4.0	59.3	7.8
C60	65	68.2	68.1	68.1	4.8	67.5	3.8
C70	75	78.2	79.3	78.7	4.9	79.0	5.3

You can also see the results for each kind of concrete in the figures 7.1 to 7.8.

Figure(7.1): Final results for C25 concrete

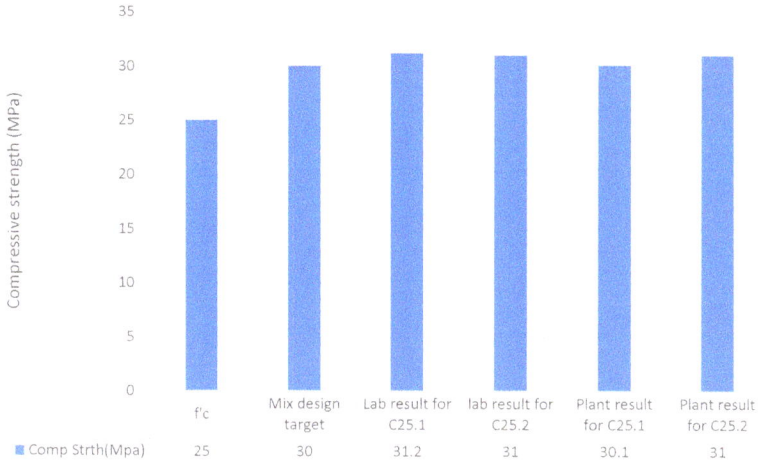

	f'c	Mix design target	Lab result for C25.1	lab result for C25.2	Plant result for C25.1	Plant result for C25.2
■ Comp Strth(Mpa)	25	30	31.2	31	30.1	31

Figure(7.2): Final results for C30 concrete

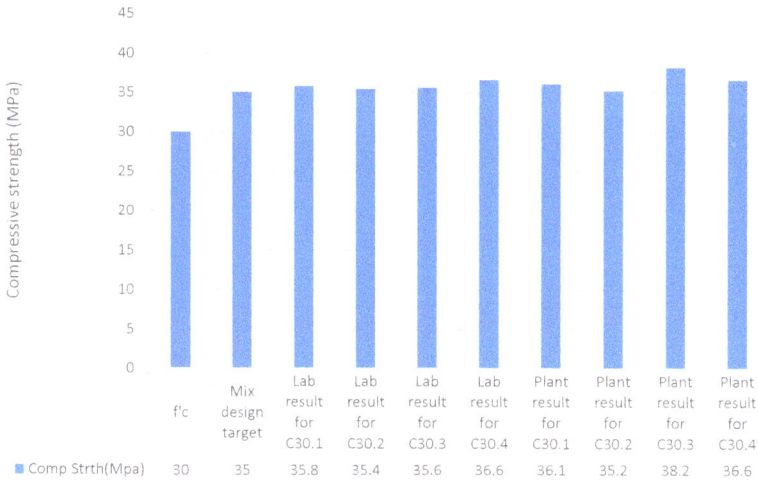

	f'c	Mix design target	Lab result for C30.1	Lab result for C30.2	Lab result for C30.3	Lab result for C30.4	Plant result for C30.1	Plant result for C30.2	Plant result for C30.3	Plant result for C30.4
■ Comp Strth(Mpa)	30	35	35.8	35.4	35.6	36.6	36.1	35.2	38.2	36.6

Figure(7.3): Final results for C35 concrete

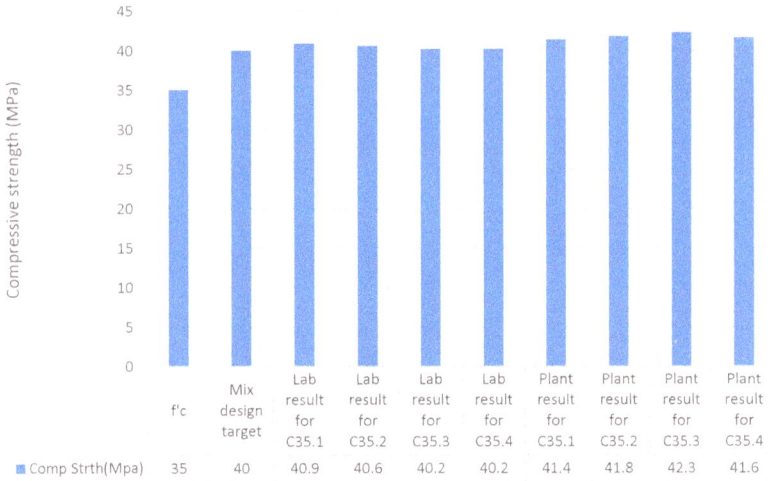

	f'c	Mix design target	Lab result for C35.1	Lab result for C35.2	Lab result for C35.3	Lab result for C35.4	Plant result for C35.1	Plant result for C35.2	Plant result for C35.3	Plant result for C35.4
Comp Strth(Mpa)	35	40	40.9	40.6	40.2	40.2	41.4	41.8	42.3	41.6

Figure(7.4): Final results for C40 concrete

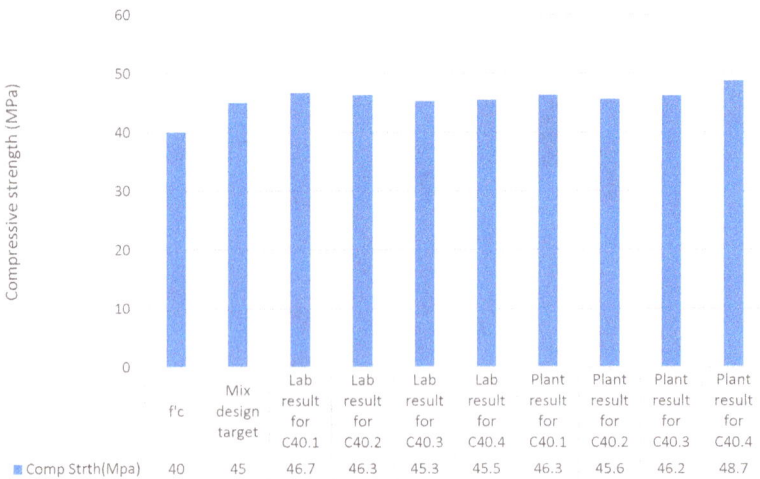

	f'c	Mix design target	Lab result for C40.1	Lab result for C40.2	Lab result for C40.3	Lab result for C40.4	Plant result for C40.1	Plant result for C40.2	Plant result for C40.3	Plant result for C40.4
Comp Strth(Mpa)	40	45	46.7	46.3	45.3	45.5	46.3	45.6	46.2	48.7

Figure(7.5): Final results for C45 concrete

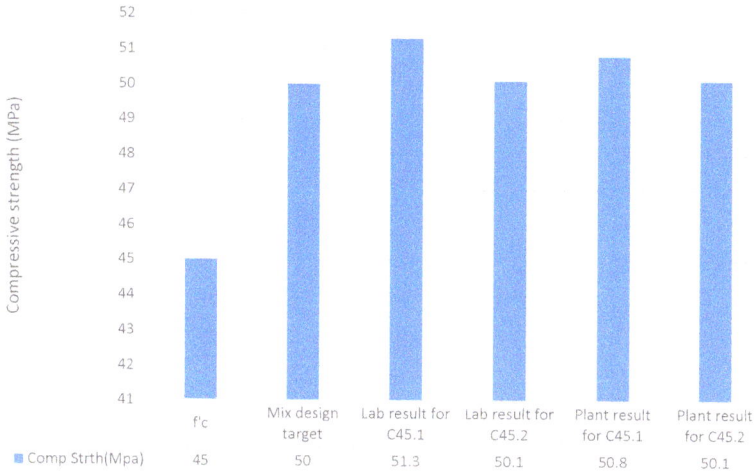

Comp Strth(Mpa)	f'c	Mix design target	Lab result for C45.1	Lab result for C45.2	Plant result for C45.1	Plant result for C45.2
	45	50	51.3	50.1	50.8	50.1

Figure(7.6): Final results for C50 concrete

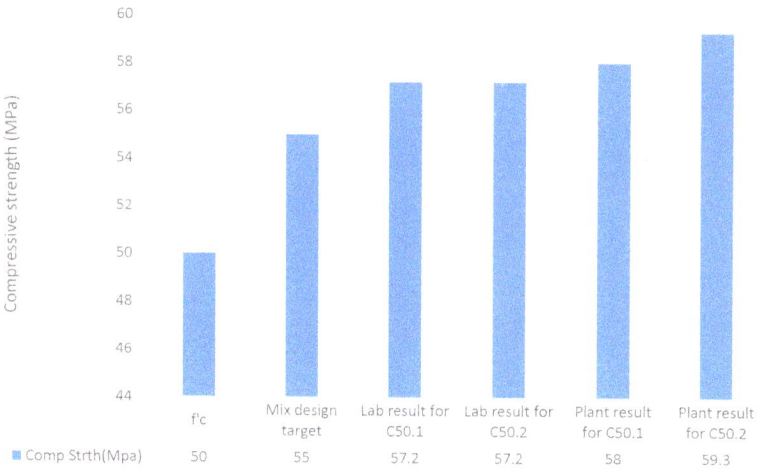

Comp Strth(Mpa)	f'c	Mix design target	Lab result for C50.1	Lab result for C50.2	Plant result for C50.1	Plant result for C50.2
	50	55	57.2	57.2	58	59.3

Figure(7.7): Final results for C60 concrete

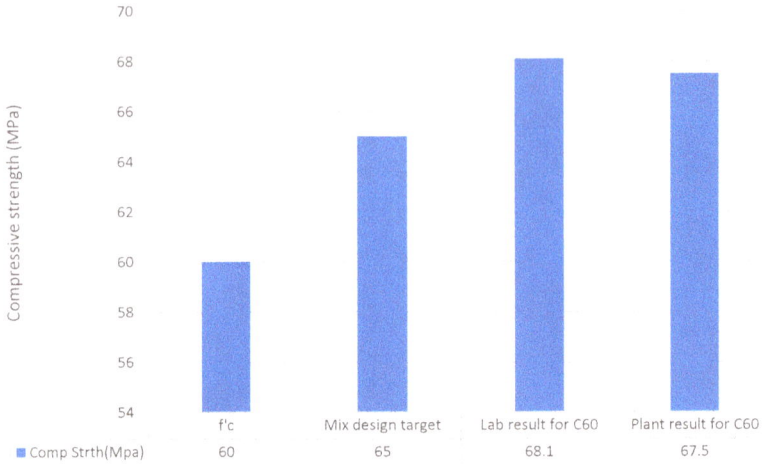

Compressive strength (MPa)	f'c	Mix design target	Lab result for C60	Plant result for C60
Comp Strth(Mpa)	60	65	68.1	67.5

Figure(7.8): Final results for C70 concrete

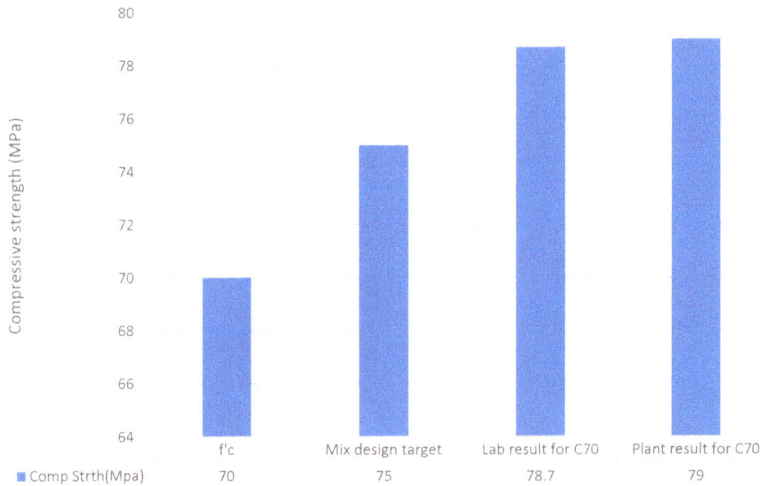

Compressive strength (MPa)	f'c	Mix design target	Lab result for C70	Plant result for C70
Comp Strth(Mpa)	70	75	78.7	79

Appendix (1):

Mini slump method

This method is in fact a slump test with a small amount of paste, using the slump cone showed in figure (a).

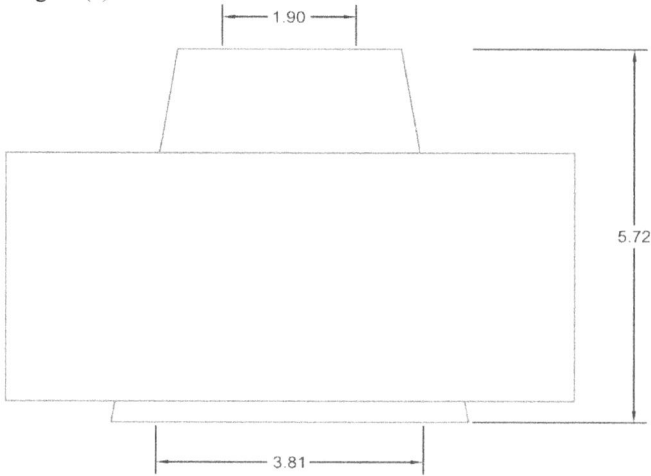

Figure (a): Mini slump cone

Preparation of the grout:

First, weigh 200 gr of cement and then water according to the w/c ratio selected. You can do the test at w /c between 0.3 to 0.45.

After weighing the super-plasticizer, add it to the water and mix it well.
Then, put the water and super-plasticizer in the cement and hand mix for one minute. Finally, mix it with an electrical mixer for about two minutes.

Testing procedure:

Put a plate on a table and check the level accurately. Then put the cone at the center of the plate and after a fifteen-second hand mixing, fill the mini cone with the paste. After five strokes, raise the cone up rapidly. Measure the diameter of the spread paste on the table along two perpendicular diameters and calculate the average of these two values.

Put the paste back to its container and cover it to avoid any desiccation and clean all of the other test vessels for the next use.

To control slump retention, you can check the slump at 10, 30, 40, 60, 90 and 120 minutes.

Figure (b): Mini slump cone

Appendix (2):

Marsh cone test

The method is to prepare a grout and to measure how long it takes for a volume of the grout to flow through a funnel by a given diameter. The cones that are used can have different diameters.

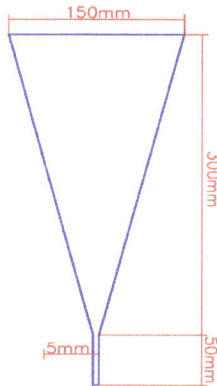

Figure (c): Marsh cone

You can see the suggested dimensions in figure (c) with the total capacity of 1.2 L. We suggest using this cone for cement super-plasticizer compatibility.

Test procedure

Do the test with a w/c ratio of 0.35 and according to this, the calculations for the amount of cement, super-plasticizer and water are done for 1.2 L.

First, weigh the water and super-plasticizer and then mix them together well. After that, weigh the cement and add it to the mixture of water and super-plasticizer progressively while mixing.

Continue hand mixing for one minute and a half. Then continue mixing for sixty seconds with an electrical mixer. Finally, let the paste flow from the marsh cone and measure the time.

To check slump retention, do the test at different times up to 90 minutes.
You can see a picture of marsh cone at figure (d).

Figure (d): Marsh cone

Appendix (3):

Material Data Sheet: BPC-40
Poly Carboxylate Ether Super-plasticizer
Product Description:
This product is a PCE base super-plasticizer to produce a very high quality flowable concrete with the most advanced technology.
Usage:
This product can reduce the amount of water in concrete production more than 35% and can be used for all kinds of concrete to: - Increase the workability of concrete - Water reduction without loss of workability - Use concrete in congested segments - Reduce human resource and vibration of concrete - Reduce permeability and increase durability of concrete - Increase compressive strength - Produce Self Consolidating Concrete(SCC)
Performance:
Poly carboxylate ether molecules separate cement particles with steric hindrance and also with electrostatic repulsion.
Product Specification:
- Appearance: Yellowish viscous liquid - Specific Gravity: 1.08 kg/L - Chloride content: <0.1% - PH : 5 to 6 - Solid Content : 40%
Standards:
This product is compatible with ASTM C494, EN 934-2
Maintenance:
To avoid quality reduction, keep between 5 to 35°C. Protect from direct sunlight.
Expiration Date:
Best to use before 12 months after the production date.
Packaging:
1000 Liter IBC tank, 20 liter and 220 Liter tanks and also bulk.
Dosage:
It can be used between 0.3 to 1.5 percent by weight of cement. To check standard compatibility, coincidence dosage is 0.6% by weight of cement.
Compatibility with cement and other admixtures:
This product is compatible with all types of cement according to ASTM and also with all kinds of blended cement or any usage of powdered additives like silica fume, GGBS, fly ash etc. This product is also compatible with most of other concrete admixtures. But you can make trials before mixing. You cannot use PCE base super-plasticizers with SNF base ones.
Curing of concrete:
Concrete which made with this kind of super-plasticizers should be cured with water or any kind of curing compounds immediately to avoid shrinkage and cracking.

Safety:	
In case of contact with skin, it can cause irritation. Wear protective gears. In the case of contact with eye and mucus tissues it should clean with enough amount of water and see a physician immediately.	
Ecology:	
Do not dispose in soil or water.	
Transportation:	
This product in nontoxic and noncorrosive. This is nonhazardous material and can be transported safely.	
Legal notes:	
The notes in this sheet is with the support of producers R&D if it is used under defined conditions. Because of probable differentiation in the raw materials, you should see the last version of data sheet in any case of use.	

References

[1]   Iranian Institute for Research on Construction Industry, 9th topic of National Rules for Construction, "Concrete Structures", 2009

[2]   Iranian National Management and Programming Organization, National Handbook of Concrete Structures, 2005

[3]   Iranian Institute for Research on Construction Industry, National Concrete Mix Design Method, 2015

[4]   American Society for Testing and Materials, Standard Test Method for Compressive Strength of Hydraulic Cement Mortars, ASTM C109-99

[5]   American Society for Testing and Materials, Standard Test Method for Density of Hydraulic Cement, ASTM C188-95

[6]   American Society for Testing and Materials, Standard Test Method for Slump of Hydraulic Cement Concrete, ASTM C143-00

[7]   American Society for Testing and Materials, Standard Test Method for Air Content of Freshly Mixed Concrete by the Pressure Method, ASTM C231-97

[8]   American Society for Testing and Materials, Standard Test Method for Compressive strength of Cylindrical Concrete Specimens, ASTM C39-01

[9]   American Society for Testing and Materials, Standard Test Method for Density, Absorption and Voids in Hardened Concrete, ASTM C642-97

[10] Aitcin P.C, High Performance Concrete, E&FN SPON, 2004

[11] Nawy G.Edward, Concrete Construction Engineering Handbook, CRC Press, 2008. https://doi.org/10.1201/9781420007657

[12] Lamond F.Joseph, Pielert H.James, Significance of Tests and Properties of Concrete and Concrete Making Materials, ASTM International, 2006. https://doi.org/10.1520/STP169D-EB

[13] European Standard Organization, Concrete-Part1: Specification, Performance, Production and Conformity, EN206-1, 2000

[14] Popovics Sandor, Concrete Materials, Properties Specification and Testing, NOYES Publications, 1992

[15] Gjorv E.Odd, Durability Design of Concrete Structures in Severe Environments, Taylor & Francis, 2009

[16] Richardson M, Fundamentals of Durable Reinforced Concrete, SPON Press, 2004

[17] Ramachandran, Paroli, Beaudion, Delgado, Handbook of Thermal Analysis of Construction Materials, NOYES Publications, 2002. https://doi.org/10.1016/B978-081551487-9.50017-7

[18] Ramachandran V.S, Concrete Admixtures Handbook, Properties, Science and Technology, NOYES Publications, 1995

[19] Ramachandran V.S, Beaudion James, Handbook of Analytical Techniques in Concrete Science and Technology, Principles, Techniques and Applications, William Andrew Publishing, 2001. https://doi.org/10.1016/B978-081551437-4.50004-7

[20] Newman John, Choo Ban Seng, Advanced Concrete Technology, Concrete Properties, Elsevier, 2003

[21] Bertolini L, Elsener B, Pedeferri P, Polder R, Corrosion of Steel in Concrete, Prevention, Diagnosis, Repair, WILEY-VCH, 2004. https://doi.org/10.1002/3527603379

[22] Iranian Standard Organization, Standard Specification for Ready Mixed Concrete, ISIRI6044, 2015

[23] Iranian Standard Organization, Concrete Admixtures, Specification, ISIRI2930, 2011

About the auhtors

Kambiz Janamian is an experienced civil engineer and concrete technologist. He worked more than 10 years in ready mixed concrete plants as the QC and development supervisor and consultant. He also worked as a concrete admixture formulator and researcher for many years. He was the supervisor for many joint projects between the concrete industry and Universities. His researches are related to concrete mix design, concrete admixtures, PCE super-plasticizers, ultra high performance concrete and many other subjects related to the concrete technology. He published 5 books in Farsi language about concrete technology with the subjects of high performance concrete, shrinkage and cracks in concrete, concrete admixtures, concrete mix design and using of plasticizers and super-plasticizers. He also translated the "SIKA Concrete Handbook" to Farsi language for the SIKA company. Currenlty, he is working for Alborz Shimie Asia, a leading company in concrete admixtures especially PCE in the middle east.

José Barroso Aguiar is an associate professor with habilitation at the Department of Civil Engineering of University of Minho, Portugal. He gained his BSc in civil engineering in 1982 and received his PhD in civil engineering in 1990. He has over 250 publications. His main areas of interest are: durability of concrete, concrete-polymer composites, incorporation of wastes in concrete and energy efficiency of buildings.

www.ingramcontent.com/pod-product-compliance
Lightning Source LLC
Chambersburg PA
CBHW071231210326
41597CB00016B/2012